# *Trichoderma* - Technology and Uses

*Edited by Fernando Cezar Juliatti*

Published in London, United Kingdom

*Trichoderma* - Technology and Uses
http://dx.doi.org/10.5772/intechopen.98115
Edited by Fernando Cezar Juliatti

Contributors
Bongani Petros Kubheka, Luwam Weldegabir Ziena, Kishor Chand Kumhar, Dalvinder Pal Singh, Anil Kumar, Amar Bahadur, Pranab Dutta, Fernando Cezar Juliatti, David de Souza Jaccoud-Filho



First published in London, United Kingdom, 2022 by IntechOpen
IntechOpen is the global imprint of INTECHOPEN LIMITED, registered in England and Wales, registration number: 11086078, 5 Princes Gate Court, London, SW7 2QJ, United Kingdom

British Library Cataloguing-in-Publication Data
A catalogue record for this book is available from the British Library

Additional hard and PDF copies can be obtained from orders@intechopen.com

*Trichoderma* - Technology and Uses
Edited by Fernando Cezar Juliatti
p. cm.
Print ISBN 978-1-80355-354-2
Online ISBN 978-1-80355-355-9
eBook (PDF) ISBN 978-1-80355-356-6

# Meet the editor

Fernando Cezar Juliatti is an agronomist with a master's degree in Phytopathology and a Ph.D. in Agronomy, Plant Breeding. He is currently a full professor at the Federal University of Uberlândia (UFU), where he has held the chair of phytopathology since 1987. He was previously director of the Institute of Agricultural Sciences and coordinator of the agronomy course at the same university. He is a member of the Chamber of Agronomy at the Amparo Research Foundation in Minas Gerais (FAPEMIG). He has published more than 220 articles in specialized journals in agronomy, plant pathology, biology, genetics, and plant breeding. He actively participates in scientific committees in Brazil, such as the Coordination of Improvement of *Higher Education Personnel Foundation* (*CAPES*), the National Council for Scientific and Technological Development (CNPq), and FAPEMIG. He twice composed the CAPES-Brazil award judging committee for the best doctoral theses in Brazil in biology, medicine, biochemistry, genetics and biotechnology, veterinary, food science, and agronomy. He has guided students awarded as winners of best works in scientific congresses and meetings. Dr. Juliatti won the Top Ethanol award from the Union of Alcohol and Sugar Industries (UNICA) in 2011. He was president of the Brazilian Society of Phytopathology in 2004 and president of the Golden Jubilee of the Brazilian Congress of Phytopathology in 2017. His research focuses on the alternative management of aggressive fungi in plants with the use of *Trichoderma*.

# Contents

# Preface

*Trichoderma* spp. is a versatile fungus with a high reproductive capacity and great potential for application in the biological control of plant diseases. It stands out for the control of phytopathogens with use via seed, soil, and straw and has action or interaction with bacteria, actinomycetes, and mycorrhizas, improving its action in hyperparasitism and antibiosis. It can also be used in the production of biofuels through the degradation of cellulose and even pharmacologically. *Trichoderma* antagonism against sclerotia producers may be imposed by different mechanisms such as mycoparasitism, antibiosis, competition, resistance induction, and plant growth promotion. The niche competition for space and nutrients besides antibiosis are the mechanisms most often used by biocontrol agents and one of the main *Trichoderma* strategies. The fast reproduction and colonization confer more effectiveness in available resources using. The successful antagonism could be attributed to the combined action of secondary metabolites and hydrolytic enzymes. The *Trichoderma* mycoparasitism relationship against pathogens may involve events such as location, recognition, direct contact, formation of hook-shaped structures with appressorium function, penetration, folding, and development of parallel hyphae. Biological control products formulated with *Trichoderma* have emerged as the best methods to manage sclerotia-producing pathogens, since chemicals in soil are often inefficient, too expensive, and too environmentally harmful. All *Trichoderma* species can utilize the cell content of other fungi as a source of nutrients. Control of white mold is also dependent on the density of sclerotia in soil. Best results are obtained when the densities are 1–10 sclerotia/kg of soil. Field experiments have shown that *Trichoderma* spp. as an exclusive control method is not sufficient to reduce the severity of white mold. In Brazil and other parts of the world, thirty-four *Trichoderma*-based products are currently registered and most of them are recommended to control soil-borne pathogens that produce sclerotia as resistance structures. Relatively few species of *Trichoderma* were developed into commercial formulations, despite the large number of publications that have shown the potential of many other species. *Trichoderma* spp. are widely used in Brazil to control white mold in soybean and its use is expected to increase, as only 50% of the infected area is currently treated with these biocontrol agents. Besides providing partial control of white mold, these fungi can also increase plant growth and productivity coupled with a reduction in the use of chemical fungicides.

**Fernando Cezar Juliatti**
Micology and Plant Protection Laboratory,
Agrarian Institute Science of University,
Federal of Uberlândia,
Minas Gerais State,
Brazil

## Section 1

# Identification of *Trichoderma* Species, Register Products and Uses in Biology Control

Chapter 1

# Introductory Chapter: *Trichoderma* the Versatile Fungus to Soil Plant Pathogens Control and Bioprocess Uses

*Fernando Cezar Juliatti and David de Souza Jaccoud-Filho*

## 1. Introduction

The *Trichoderma* mycoparasitism relationship against pathogens may involve events such as location, recognition, direct contact, formation of hook-shaped structures with appressorium function, penetration, folding, and development of parallel hyphae [1–4]. Some studies report that *Trichodema* species are efficient in antagonizing phytopathogens with resistance structures considered difficult to be attacked by microorganisms, such as spores, sclerotia, chlamydospores, and microsclerotia [5, 6].

*Trichoderma* antagonism against sclerotia producers may be imposed by different mechanisms, such as mycoparasitism, antibiosis, competition, resistance induction, and plant growth promotion [1, 7–9]. The niche competition for space and nutrients besides antibiosis are the most often mechanisms used by biocontrol agents, and one of the main *Trichoderma* strategies. The fast reproduction and colonization confer more effectiveness on available resources using. By the way, the successful antagonism could be attributed to the combined action of secondary metabolites and hydrolytic enzymes [10]. The wide range of secondary metabolites includes epipolythiodioxopiperazines (ETPs), peptaibols, pyrones, butenolides, pyridines, azaphilones, steroids, anthraquinones, lactones, trichothecenes, and harzianic acid [8]. These metabolites can interfere with the metabolic activities of other microorganisms, promoting growth and sporulation inhibition, reduction in spore germination, in addition to hyphae distortions, and endolysis. Some *Trichoderma* species are strong cellulases, chitinases, and β-1,3-glucanases producers. These enzymes are involved in the fungi and oomycetes cell wall components degradation process and can interfere in its biosynthesis [2, 11–15]. Proteases and lipases can also kill some fungi, which are substrate for mycoparasites [5, 16]. Different enzyme cell wall degradation-related are expressed in *Trichoderma harzianum* in biocontrol when grown on mycelia, sclerotia, or apothecia of *S. sclerotiorum*. There is probably a synergistic action between the cell wall-degrading enzymes [2].

The *Trichoderma* mycoparasitism relationship against pathogens may involve events such as location, recognition, direct contact, formation of hook-shaped structures with appressorium function, penetration, folding, and development of parallel hyphae [1–4, 6, 17–19]. Some studies report that *Trichodema* species are efficient in antagonizing phytopathogens with resistance structures considered difficult to be attacked by microorganisms, such as spores, sclerotia, chlamydospores, and microsclerotia [5, 6, 17]. Besides that, *Trichoderma* application in seeds improves the seed sanity, physiological quality, germination, and early development, as observed for soybean seeds [9].

## 2. Biological control by *Trichoderma*

In 2014 there were 177 *Trichoderma*-based fungicides commercially available in the world [20]. These products contained mainly *Trichoderma asperellum, T. hamatum, T. harzianum,* and *T. viride* as active ingredients and were recommended mainly for seed and soil treatments [20]. In Brazil, there are currently 34 formulated products with *Trichoderma* as active ingredients registered in the Ministry of Agriculture, Livestock and Food Supply (MAPA) [21, 22]. These 34 products are based on four species: *T. harzianum*, *T. asperellum*, *T. koningiopsis* and *T. stromaticum*. The combination of more than one biocontrol agent is thought to be advantageous, but it depends on the individual strains' compatibility [23]. Six out of the 34 registered products in Brazil are formulated with one or two *Trichoderma* and *Bacillus amyloliquefaciens* strains.

However, it is not known whether these microorganisms are compatible or not or if there is any synergism in their combination. Bacterial genera such as *Bacillus* and *Pseudomonas* are potential biocontrol agents of soil-borne pathogens due to the secretion of antibiotics and lytic enzymes in the rhizosphere of plants. Therefore, they are potential agents to be combined with *Trichoderma*, especially when they do not inhibit each other [23–27]. However, the compatibility of combinations needs to be evaluated with *in vitro* and *in planta* assays [28]. Interactions between *Trichoderma* and mycorrhyzae are sometimes antagonistic, such as with the ectomycorrhyzal basidiomycetous genus *Laccaria* spp., where there was clear inhibition of growth, colonization, and spore germination on both partners [29–31]. However, sometimes these interactions are synergistic, such as with *Glomus* spp. Although there was an increase in plant biomass in the interaction, microscopical observations clearly showed that *Trichoderma* was parasitizing this endomycorrhyzal fungus [32].

The fungicides thiophanate methyl+fluazinam, carbendazim+tiram, metalaxyl-M+fludioxonil, fipronil+pyraclostrobin+thiophanate-methyl and carboxine+tiram are compatible with the isolates of *Trichoderma* spp. The fungicide carboxin+tiram is compatible with the isolates at high concentration of 1000 ppm, thus suggesting that these fungicides can be used in the integrated management with the isolates *Trichoderma* spp. [9].

## 3. Conclusion

*Trichoderma* spp. is a versatile fungus with a high reproductive capacity and with a high potential for application in the biological control of plant diseases. It stands out for the control of phytopathogens with use via seed, soil, straw, and has action or interaction with bacteria, actinomycetes, and mycorrhizas, improving its action in hyperparasitism and antibiosis. It can also be used in the production of biofuels, through the degradation of cellulose, and even pharmacologically.

## Acknowledgements

FCJ acknowledges CNPq (National Council for Scientific and Technological Development) for his productivity scholarship. DSJF acknowledges Fundação Araucária (Araucária Foundation to Support Scientific and Technological Development of Paraná) for his productivity scholarship.

## Conflict of interest

The authors declare no conflict of interest.

## Author details

Fernando Cezar Juliatti[1*] and David de Souza Jaccoud-Filho[2]

1 Micology and Plant Protection Laboratory, Agrarian Institute Science of University, Federal of Uberlândia, Brazil

2 Applied Plant Pathology Laboratory, UEPG, Ponta Grossa, Brazil

*Address all correspondence to: juliatti@ufu.br

## References


[1] Zhang F, Ge H, Zhang F, Guo N, Wang Y, Chen L, et al. Biocontrol potential of *Trichoderma harzianum* isolate T-aloe against *Sclerotinia sclerotiorum* in soybean. Plant Physiology and Biochemistry. 2016;**100**:64-74. DOI: 10.1016/j.plaphy.2015.12.017

[2] Troian RF, Steindorff AS, Ramada MHS, Arruda W, Cirano JU. Mycoparasitism studies of *Trichoderma harzianum* against *Sclerotinia sclerotiorum*: Evaluation of antagonism and expression of cell wall-degrading enzymes genes. Biotechnology Letters. 2014;**36**:2095-2101. DOI: 10.1007/s10529-014-1583-5

[3] Abdullah MT, Ali NY, Suleman P. Biological Control of *Sclerotinia sclerotiorum* (Lib.) de Bary with *Trichoderma harzianum* and *Bacillus amyloliquefaciens*. Crop Protection. 2008;**27**:1354-1359. DOI: 10.1016/j.cropro.2008.05.007

[4] Silva FF, Castro EM, Moreira SI, Ferreira TC, Lima AE, Alves. Emergência e análise ultraestrutural de plântulas de soja inoculadas com *Sclerotinia sclerotiorum* sob efeito da aplicação de *Trichoderma harzianum*. Summa Phytopathologica. 2017;**43**:41-45. DOI: 10.1590/0100-5405/2212

[5] Melo IS. Potencialidades da utilização de *Trichoderma* spp. no controle biológico de doenças de plantas. In: Bettiol W, editor. Controle biológico de doenças de plantas. Jaguariúna: CNPDA/EMBRAPA; 1991. pp. 135-156

[6] Marcello CM. Avaliação da expressão e caracterização de uma exo-β-1,3-glucanase envolvida no mecanismo de micoparasitismo de *Trichoderma asperellum* [thesis]. Brasília: Universidade de Brasília; 2008

[7] Adnan M, Islam W, Shabbir A, Khan KA, Ghramh HA, Huang Z, et al. Plant defense against fungal pathogens by antagonistic fungi with *Trichoderma* in focus. Microbial Pathogenesis. 2019;**129**:7-18. DOI: 10.1016/j.micpath.2019.01.042

[8] Khan RA, Najeeb S, Hussain S, Xie B, Li Y. Bioactive secondary metabolites from *Trichoderma* spp. Against phytopathogenic fungi. MDPI Microorganisms. 2020;**8**:817. DOI: 10.3390/microorganisms8060817

[9] Juliatti FC, Rezende AA, Juliatti BC, Morais TP. *Trichoderma* as a Biocontrol Agent against *Sclerotinia* Stem Rot or White Mold on Soybeans in Brazil: Usage and Technology. In: Shar MM, editor. Trichoderma: The Most Widely Used Fungicide. London: InTech Open; 2019. pp. 1-23. DOI: 10.5772/intechopen.84544

[10] Druzhinina IS, Seidl-Seiboth V, Herrera-Estrella A, Horwitz BA, Kenerley CM, Monte E, et al. *Trichoderma*: The genomics of opportunistic success. Nature Review Microbiology. 2011;**9**:749-759. DOI: 10.1038/nrmicro2637

[11] Bettiol W, Morandi MAB, Pinto ZV, Paula Junior TJ, Correa EB, Lucon CMM, et al. Produtos Comerciais à Base de Agentes de Biocontrole de Doenças de Plantas. Jaguariúna: Embrapa Meio Ambiente; 2012. p. 155

[12] Dias PP. Controle biológico de fitopatógenos de solo por meio de isolados de fungos do gênero *Trichoderma* e sua contribuição no crescimento de plantas. 2011. 101 f. [thesis]. Seropédica: Universidade Federal Rural do Rio de Janeiro; 2011

[13] Amorim L, Rezende JAM, Bergamin Filho A, editors. Manual de Fitopatologia vol. I—Princípios e Conceitos 5th ed. Ouro Fino: Agronômica Ceres; 2018. p. 810

[14] Carvalho Filho MR. Relações filogenéticas, identificação e potencial de uso de isolados de *Trichoderma* no controle do mofo branco e como promotores de crescimento do feijoeiro [thesis]. Brasília: Universidade de Brasília; 2013

[15] Souza JR. Potencialidade de fungicida e agente biológico no controle de requeima do tomateiro [dissertation]. Vitória da Conquista: Universidade Estadual do Sudoeste da Bahia; 2013

[16] Harman GE. The Nature and Application of Biocontrol Microbes II: *Trichoderma* spp.—Overview of Mechanisms and Uses of *Trichoderma* spp. Phytopathology. 2005;**96**:190-194. DOI: 10.1094/PHYTO-96-0190

[17] Souza ACA, Sousa TP, Cortês MVB, Rodrigues FA, Silva GB, et al. Enzyme induced defense response in the suppression of rice leaf blast (*Magnaporthe oryzae*) by silicon fertilization and bioagents. International Journal of Research Studies in Bioscience. 2015;**3**:22-32

[18] Bozzola JJ, Russell LD. Electron Microscopy. Boston: Jones and Bartlett Publishers; 1999. p. 670

[19] Alves E, Lucas GC, Pozza EA, Carvalho AM. Scanning electron microscopy for fungal sample examination. In: Laboratory Protocols in Fungal Biology. New York: Springer; 2013. pp. 133-150

[20] Woo SL, Ruocco M, Vinale F, Nigro M, Marra R, Lombardi N, et al. *Trichoderma*-based Products and their Widespread Use in Agriculture. The Open Mycology Journal. 2014;**8**:271-126

[21] MAPA. Ministério da Agricultura Pecuária e Abastecimento. Agrofit—Sistemas de Agrotóxicos Fitossanitários, Coordenação Geral de Agrotóxicos e Afins. [Internet]. 2021. Available from: http://agrofit.agricultura.gov.br/agrofit_cons/principal_agrofit_cons. [Accessed: February 19, 2022]

[22] Lucon CMM, Chaves ALR, Bacilieri S, editors. *Trichoderma*: o que é, para que serve e como usar corretamente na lavoura. São Paulo: Instituto Biológico; 2014. p. 28

[23] Barbosa LO, Lima JS, Magalhaes VC, Gava CAT, Soares ACF, Marbach PAS, et al. Compatibility and combination of selected bacterial antagonists in the biocontrol of sisal bole rot disease. BioControl. 2018;**63**:595-605. DOI: 10.1007/ s10526-018-9872-x

[24] Mishra J, Arora NK. Secondary metabolites of fluorescent pseudomonads in biocontrol of phytopathogens for sustainable agriculture. Applied Soil Ecology. 2017;**125**:35-45. DOI: 10.1016/j.apsoil.2017.12.004

[25] Pawana G, Badami K, Djunaedy A. Growth and Reproduction of *Trichoderma* sp. in with presence *Bacillus* sp. or Fluorescent Pseudomonad. Journal of Physics: Conference Series. 2021;**1899**:1-7. DOI: 10.1088/1742-6596/1899/1/012021

[26] Pan S, Jash S. Variability in biocontrol potential and microbial interaction of *Trichoderma* spp. with soil inhabiting antagonistic bacteria *Pseudomonas fluorescens*. Indian Phytopathology. 2010;**63**:158-164

[27] Braga AF. Interaction of *Trichoderma asperellum* and biological control of

*Bacillus* spp. used in soybean diseases [dissertation]. Rio Verde: Instituto Federal Goiano; 2021

[28] Cruz-Magalhães V, Guimarães RA, Silva JCP, Faria AF, Pedroso MP, Campos VP, et al. The combination of two Bacillus strains suppresses *Meloidogyne incognita* and fungal pathogens, but does not enhance plant growth. Pest Managment Science. 2021;**78**:722-732. DOI: 10.1002/ps.6685

[29] Summerbell RC. The inhibitory effect of *Trichoderma* species and other soil microfungi on formation of mycorrhiza by *Laccaria bicolor* in vitro. New Phytology. 1987;**105**:437e448. DOI: 10.1111/j.1469-8137.1987.tb00881.x

[30] Werner A, Zadworny M, Idzikowska K. Interaction between *Laccaria laccata* and *Trichoderma virens* in co-culture and in the rhizosphere of *Pinus sylvestris* grown *in vitro*. Mycorrhiza. 2002;**12**:139e145. DOI: 10.1007/s00572-002-0159-8

[31] Guo Y, Ghirardo A, Weber B. *Trichoderma* species differ in their volatile profiles and in antagonism toward ectomycorrhiza *Laccaria bicolor*. Frontiers in Microbiology. 2019;**10**:1-15. DOI: 10.3389/fmicb.2019.00891

[32] Rousseau A, Benhamou N, Chet I, Piche Y. Mycoparasitism of the extramatrical phase of *Glomus intraradices* by *Trichoderma harzianum*. Phytopathology. 1996;**86**:434-443. DOI: 10.1094/Phyto-86-434

Chapter 2

# *Trichoderma*: A Biofertilizer and a Bio-Fungicide for Sustainable Crop Production

*Bongani Petros Kubheka and Luwam Weldegabir Ziena*

## Abstract

*Trichoderma* has been studied widely. It has been found to play a major role in agricultural production. Around the world scientists and farmers have taken advantage of this knowledge. It is reported to improve plant growth of many crops such as tomato, lettuce, maize, beans, cabbage sugarcane and many more crops. There are two broad categories where *Trichoderma* plays a major role which is its use as a biofertilizer as well as a biofungicide. Its use as a biofertilizer has been aggravated by its ability to produce volatile compounds, ability to solubilize phosphates making them available to the plant. Moreover, farmers use it as a biofertilizer because it improves the uptake of macro and micro nutrients by the plant. As a biofungicide, *Trichoderma* is not to control many pathogens from various crops. This includes the control of pathogens such as *Rhizoctonia, Phytophthora, Rhizoctonia, Sclerotinia, Phythium, Fusarium, Sclerotinia* species *and Galumannomyces*. The mechanisms used by *Trichoderma* as a biofungicide includes, antibiosis, mycoparasitism, competitive advantage in the rhizosphere as well as priming of the crop self-defense mechanisms. The purpose of this book chapter is to highlight the importance of *Trichoderma* in agriculture as a biofertilizer and biofungicide.

**Keywords:** biofertilizer, biofungicide, phytohormones, volatile compounds, phosphates, nutrient uptake, antibiosis, mycoparasitism, competition, resistance

## 1. Introduction

The increase in the human population around the world has pushed farmers to produce more food. This pressure forced some farmers to use more chemicals in their operations, which led to concerns raised by environmentalists and health officials as some chemicals were damaging the environment and people's health. This has raised a necessity of exploring alternative methods to improve fertilization, and manage pests and diseases.

Biofertilizer became an option as it is friendlier to the environment as well as on human health. *Trichoderma* is one of the fungal cultures that has been studied for this purpose [1]. It has been found that *Trichoderma* can produce various plant growth-promoting compounds such as enzymes and phytohormones [2]. Some enzymes produced, helps the plant to access nutrients that are not accessible by the plant due to their form. For example acid soils tend to bind phosphorus forming toxic complexes rendering the phosphorus unavailable to the crop. This results in the crop not getting the nutrients that were meant for it, thus reducing the crop yields. Some enzymes produced by *Trichoderma* solubilize phosphates making the phosphorus available again to the crop [3].

Phytohormones on the other hand are compounds that are responsible for the growth and development of the plant. Some are responsible for plant elongation, shoot and root developments, others are involved in plant pests and disease control [4]. *Trichoderma* has been reported to produce some of the plant growth hormones such as indole-3-acetic acid (IAA), Auxins, gibberellic acid [5–7]. Scientists and farmers exploit these properties by developing biofertilizers, in this case using *Trichoderma* as the organism that can produce multiple phytohormones [8].

Biofungicides also became an alternative to chemical or synthetic fungicides to minimize the damage caused by chemical fungicides to the environment, animals and human beings. *Trichoderma* is one of the fungi that is also used as biofungicides by farmers as it has various mechanisms for controlling growth and development of several plant pathogens. *Trichoderma* is reported to produce antibiotics [9, 10], volatile compounds [11, 12], induce or prime plant resistance [13]. Moreover, it is reported to compete better than other microorganisms in the rhizosphere and has mycoparasitism behavior [14].

The objective of this book chapter is to prove that *Trichoderma* may be used as both a biofertilizer and biofungicide providing a sustainable alternative to chemical methods of fertilization and controlling plant pathogens.

## 2. *Trichoderma* as a biofertilizer

*Trichoderma* has been reported to promote plant growth in various ways. Some people have used it as a biofertilizer because of its ability to stimulate plant growth of many crops. It comes as an alternative to chemical fertilizer or as an amendment to improve crop production. Many attributes qualifies it to be used as an alternative or amendment to improve fertilization sustainably. Some of them are the following facts:

- It produces plant growth hormones and volatile compounds;
- It contributes to solubilizing phosphates that are unavailable to the crop
- It also takes part in promoting the uptake of macro and micro nutrients needed by the crop

### 2.1 Production of plant growth hormones and volatile compounds

Plant growth hormones are also called phytohormones. They are involved in many processes in the plant including communication, biotic and abiotic stress management, and many more processes. They have been reported for many years to play a vital role in the growth and development of a crop. Root and shoot elongation needs phyto-hormones to happen properly at the correct speed that supports high productivity. It has been reported that the presence of *Trichoderma* increases the production of some growth hormones such as indole-3-acetic acid (IAA) and gibberellic acid [15]. These two hormones are important in promoting plant growth, they are responsible for plant elongation [7]. As stated in **Table 1**, *Trichoderma* also improves germination rate and improves seedling vigor, which is an advantage for the crop. This is also associated with balanced phytohormones.

| ***Trichoderma* strain** | **Intended use** | **Target crop** | **Mode of application** | **Benefits/comments** | **Ref. (s)** |
|---|---|---|---|---|---|
| *Trichoderma azevedoi* | Growth promotion and inhibition of phytopathogen development | Lettuce | Expose plants to *T. azevedoi* | Increased chlorophyll content, and carotenoids. Decreased the severity of white mold by up to 78.83% | [16] |
| *Trichoderma afroharzianum* | Growth promotion | Tomato | Seed treatment | Phytohormone homeostasis, antioxidant activity, phenylpropanoid biosynthesis and glutathione metabolism | [17] |
| *Trichoderma harzianum, Trichoderma asperellum, Trichoderma hamatum, and Trichoderma atroviride* | Biofertilizer | Chinese cabbage | Through irrigation | Increased yield by 37%; Increased enzyme activity in the soils (urease by 25.1%, phosphatase by 13.1%, and catalase by 14.0%, Providing more inorganic nitrogen and phosphorus to the soil | [18] |
| *T. asperellum* strain GDFS1009 | Soil conditioner | Maize | On soil as granules | Increased yields | [19] |
| *Trichoderma brevicompactum, Trichoderma gamsii and T. harzianum* | Biofertilizer | Tomato | Seedling drenching | Produces indole-3 acetic acid and | [15] |

**Table 1.**
*Production of plant growth hormones and volatile compounds by Trichoderma to improve plant growth.*

## 2.2 Solubilization of phosphates

Phosphorus is one of the critical nutrients that plants need for their growth and development. It is found in the soil but due to depletion farmers have to apply fertilizers. However, the availability of phosphorus to the crop depends on the form it is in. Acidic soils bind phosphorus and make it unavailable to the crop, which is an undesired outcome [20]. Due to this, the accuracy of the amount required by the crop may not be achieved resulting in challenges associated with lack or insufficient phosphorus in the soil [3]. Some microorganisms mediate this process by solubilizing phosphates, converting them back to be in the available form for crop utilization. *Trichoderma* species have been reported to be one of those organisms. Species such as *Trichoderma harzianum* [21], *Trichoderma reesei* [22], solubilize phosphates through the production of enzymes called phytase. The phytase activity is induced by the presence of insoluble tricalcium phosphate [5]. Other *Trichoderma* species such as *Trichoderma koningiopsis* solubilize phosphates by producing alkaline phosphatase enzymes (**Table 2**) [6].

| *Trichoderma* strain | Intended use | Target crop | Mode of application | Benefits/comments | Ref. (s) |
|---|---|---|---|---|---|
| *T. harzianum, T. asperellum, T. hamatum, and T. atroviride* | Biofertilizer | Chinese cabbage | Through irrigation | Increased yield by 37%; Increased enzyme activity in the soils (urease by 25.1%, phosphatase by 13.1%, and catalase by 14.0%, Providing more inorganic nitrogen and phosphorus to the soil | [18] |
| *T. harzianum Rifai; T. asperellum T42* | Biofertilizer | Tomato | Seed treatment | Increase Phosphorus uptake | [23] |
| *T. brevicompactum, T. gamsii and T. harzianum* | Biofertilizer | Tomato | Seedling drenching | Phosphorus solubilization | [15] |

**Table 2.**
*Solubilization of phosphates by Trichoderma to promote plant growth.*

## 2.3 Macro and micro nutrient uptake

It has been reported that plant nutrient uptake can be improved resulting in plant growth promotion. Microorganisms play a major role in accelerating nutrient uptake. *Trichoderma* is one of those microorganisms that contribute to nutrient uptake [24].

| *Trichoderma* strain | Intended use | Target crop | Mode of application | Benefits/comments | Ref. (s) |
|---|---|---|---|---|---|
| *T. azevedoi* | Growth promotion and inhibition of phytopathogen development | Lettuce | Expose plants to *T. azevedoi* | Increased the content of chlorophyll, and carotenoids. Decreased the severity of white mold by up to 78.83% | [16] |
| *Trichoderma erinaceum* | Biofertilizer and biofungicide | Rice | Seed treatment | Improved germination rate and enhanced vigor. Increased yields | [27] |
| *T. harzianum T22* | Biofertilizer | Tomato | Seed treatment | Improved soil fertility, nutrient uptake, increased yields, antioxidants and minerals | [28] |
| *T. harzianum, T. asperellum, T. hamatum, and T. atroviride* | Biofertilizer | Chinese cabbage | Through irrigation | Increased yield by 37%; Increased enzyme activity in the soils (urease by 25.1%, phosphatase by 13.1%, and catalase by 4.0%, Providing more inorganic nitrogen and phosphorus to the soil | [18] |

| ***Trichoderma* strain** | **Intended use** | **Target crop** | **Mode of application** | **Benefits/comments** | **Ref. (s)** |
|---|---|---|---|---|---|
| *T. harzianum Rifai; T. asperellum T42* | Biofertilizer | Tomato | Seed treatment | Improves nutrient uptake (enhance nitrogen utilization efficiency, increase Phosphorus uptake | [23] |
| *T. harzianum T22* | BioF/compost | Tomato | Soil amendment compost | Provided 12.9% yield increase compared to recommended fertilization | [29] |
| *T. asperellum strain* GDFS1009 | Soil conditioner | Maize | On soil as granules | Increased yields | [19] |
| *T. viride* | Biofertilizer | Sugar cane | Powder broadcasted with fertilizer | Improve nutrient uptake NPK | [25] |
| *T. asperellum T34* | Biofertilizer -micronutrient | Cucumber | Seedling drenching | Enhance Fe and Cu uptake by plants | [30] |
| *T. harzianum* | Compost | Most crops | Compost | Improves the rate of Residue decomposition resulting in greater availability of soil nutrients | [31] |
| *Trichoderma simmonsii* | Biofertilizer | Bell pepper | Seedling drenching | Bell pepper yield increase up to 67%. Enhance tolerance to abiotic stresses | [32] |
| *T.brevicompactum, T. gamsii and T.harzianum* | Biofertilizer | Tomato | Seedling drenching | Improve nutrient uptake | [15] |
| *Trichoderma harzianum T22* | Biofertilizer BioF/compost | Tomato | Seed treatment | Improved soil fertility, nutrient uptake, increased yields, antioxidants and minerals (P, K, Ca, Mg, Cu, Fe, Mn and Zn) | [28, 29] |
| *T. asperellum T34* | Biofertilizer -micronutrient | Cucumber | Seedling drenching | Enhance Fe and Cu uptake by plants | [30] |
| *T. harzianum* | Compost | Most crops | Compost | Improves the rate of Residue decomposition resulting in greater availability of soil nutrients | [31] |
| *T. reesei* | Plant growth promoter | Chickpea | Seed treatment | mineral mobilization and their uptake | [33] |

**Table 3.**
*Trichoderma improving macro and micro nutrient uptake by crops.*

In a sugarcane study, there was an increase in nitrogen, potassium, phosphorus and organic carbon after the inoculation with *Trichoderma viride* [25]. Nutrient availability, as well as uptake, is improved by the presence of *Trichoderma* in the rhizosphere. The nutrient uptake is improved because of the conversion of the required nutrients from being unavailable to the plant to an available form. For example in acidic soils the applied chemical fertilizer is converted to an unavailable form to the plant, forming complexes that may be even toxic to the plant such as aluminum complexes [26]. Its the ability to colonize roots well that gives it an advantage over other microorganisms and enables the crop to receive more from it than others. Therefore, it provides a better and sustainable fertilization as it will be present in the root system as endophytes as well as root colonizers for a longer time than chemical fertilizers. Chemical fertilizers get used up as they do not multiply as microorganisms do. Sustainability is one of the potential benefits that *Trichoderma* provides as a biofertilizer. Other farmers apply microorganisms to improve fertilizer use efficiency by mobilizing nutrients that have accumulated in the soils yet are not available to the plant (**Table 3**) [34].

## 3. *Trichoderma* as a biofungicide

Agriculture is an indispensable part of any country to feed the millions of people. However, production is hampered by various plant diseases posing serious yield reductions threatening global food security. Disease management employs mainly synthetic fungicides. However, with the mounting concern for human health and environmental risks, and the loss of pesticides to resistance, the search for non-chemical alternatives has been a focus of much research for more than three decades. Biocontrol agents have emerged as an important component of plant disease management, and may provide an alternative to synthetic fungicides.

*Trichoderma* species, free-living and cosmopolitan fungi found abundantly in the soil, decaying organic and vegetable matter, were first reported as biocontrol agents in the early 1930s in the control of root rot causing *Armillaria mellea* in citrus [35].

They are successful antagonists having biocontrol abilities against a broad range of economically important phytopathogenic fungi such as *Phytophthora, Rhizoctonia, Sclerotium, Phythium, Fusarium, Sclerotinia, and Galumannomyces. Trichoderma harzianum, Trichoderma viride and Trichoderma koningii* are the main species viz. presently mass-produced by entrepreneurs [36–40].

*Trichoderma* species have been of particular interest as biocontrol as due to their rapid growth and capability of utilizing different substrates, species of this genus are often predominant components of the soil mycoflora in various ecosystems. Their ability to produce hydrolytic enzymes, secondary metabolites and degradation of xenobiotics is also an additional advantage that have an important economic impact [31, 41–43].

Competition for nutrient and ecological niche, mycoparasitism and antibiosis are the major biological mechanisms involved in their direct antagonistic activity against plant pathogenic fungi [43–45]. They can also achieve an indirect effect of antagonism on the target pathogen by interacting with the host tissue, inducing host resistance which protects against the pathogen, promoting plant and root growth as well as improving plant stress tolerance. Many successful biocontrol agents use a combination of different modes of action to produce a higher level of antagonism [38, 46].

### 3.1 Antibiosis as a mechanism of pathogen control

Antibiosis involves the production of various antimicrobial compounds by *Trichoderma* strains that inhibit or reduce the growth and/or proliferation of phytopathogens [44]. Most *Trichoderma* strains also produce volatile and non-volatile toxic metabolites that inhibit colonization by antagonized microorganisms; among these metabolites, the production of harzianic acid, alamethicins, tricholin, peptaibols, antibiotics, 6-penthyl-a-pyrone, massoilactone, viridin, gliovirin, glisoprenins, heptelidic acid and others have been described [47–50]. This phenomenon has been observed in various fungi including *Trichoderma*, which can produce a multitude of compounds with antagonistic properties including cell wall degrading enzymes such as cellulase, xylanase, pectinase, glucanase, lipase, amylase, arabinase, and protease, volatile metabolites such as 6-n-pentyl-2H-pyran-2-one (6-PAP) [51–53], and several antibiotics such as trichodermin, trichodermol, gliovirin, gliotoxin, viridin, herzianolide, pyrones, peptaibols, ethylene and formic aldehyde [50, 54, 55]. In general, strains of *T. virens* with the best efficiency as biocontrol agents can produce gliovirin [50].

### 3.2 Mycoparasitism

Mycoparasitism, direct contact of an antagonist with a fungal pathogen, involves sequential events, including pathogen recognition, attack and subsequent penetration of the host cell and death [10]. In this process, *Trichoderma* species initially produce cell wall degrading enzymes at low levels in an attempt to identify its prey. Upon recognition, growth towards the direction of the target pathogen area is induced together with a higher production of cell wall degrading enzymes (CWDEs), mainly chitinases, glucanases and proteases [56, 57]. *Trichoderma* species will then attach to their prey by binding to the carbohydrates present in the *Trichoderma* to the lectins of the fungi, followed by coiling around the pathogen's hyphae and appressoria development to penetrate the hyphae, which are subsequently attacked and degraded through the production of hydrolytic enzymes and secondary metabolites. Other CWDEs constituting hydrolysing polymers such as β-1,6-glucans and α-1,3-glucans are reported to further ensure complete disintegration of fungal mycelia or conidia [43, 58].

### 3.3 Competition in the rhizosphere

Starvation is the most common cause of death for microorganisms, so the limited availability of and competition for micro- and macro nutrients results in the biological control of fungal phytopathogens [59]. *Trichoderma* exhibits a better capability of absorption and mobilization of nutrients from the soil in comparison to other rhizospheric microorganisms; therefore, the biocontrol of fungal pathogens using *Trichoderma* involves the coordination of numerous strategies, such as the competition for nutrients, which is considered among the most important [60, 61]. In most filamentous fungi, iron uptake is essential for viability, and under iron starvation, most fungi excrete low molecular-weight ferric-iron specific chelators, termed siderophores, to mobilize environmental iron [62]. Certain *Trichoderma* strains can produce siderophores by trapping the ferric ions from the shared niche inhibiting the growth and activity of soil-borne fungal pathogens such *Botrytis cinerea* [63].

### 3.4 Priming of resistance mechanism in host plants

During plant–pathogen interactions, plants have evolved a wide range of defense mechanisms to cope with the constant attack by invading pathogens. However, plant defense can also be triggered by biocontrol agents [2, 54]. The rhizocompetent nature of *Trichoderma* species allows it to colonize roots, triggering the plant immune system (induced systemic resistance; ISR), and pre-activation (priming) of the molecular mechanisms of defense against several

| *Trichoderma* strain | Disease | Mode of action | Ref. (s) |
|---|---|---|---|
| *Trichoderma pseudokoningii* | *Fusarium oxysporum* | Antibiosis metacaspase-independent Apoptotic cell death. | [74, 75] |
| *T. harzianum* | *B. cinerea; F. oxysporum Sclerotium rolfsii* | Endo-chitinase, chitobiosidase | [63, 76, 77] |
| *Trichoderma virens* | *Rhizoctonia solani* | Colonization and antibiosis | [10, 48, 78] |
| *T. brevicrasum* | *R. solani* | Mycoparasitism | [14] |
| *T. harzianum* | *B. cinerea* | Competition for space | [45] |
| *T. asperellum* | *Pseudomonas syringae* | Induced resistance | [69] |

**Table 4.**
*Examples of Trichoderma antagonists used for successful control of fungal diseases and possible mode of action.*

potent phytopathogens and the stress challenged conditions [64–67]. Furthermore, colonization of this beneficial fungi promotes plant growth and also upgrades the host plants against various abiotic and biotic stresses [7, 68]. It balances different phytohormone-dependent pathways among which salicylic acid (SA), jasmonates (JA), ethylene (ET), abscisic acid (ABA), auxin (indole-3-acetic acid: IAA), and gibberellins (GA) are the most relevant—and modulating the levels of growth and defense regulatory proteins [2, 11, 54, 69–71]. Priming facilitates a faster and stronger reaction if the stress recurs. Reinforced responses to pathogen attacks come under the category of induced defense, while responses to abiotic are referred to as acclimation or hardening, even though these responses are similar at the beginning. They can also be enhanced by priming treatments [72, 73]. An accurate definition of how *Trichoderma* exerts its beneficial action on plants is of particular relevance to the way in which commercial products based on the abilities of *Trichoderma* are registered (**Table 4**) [15, 79, 80].

## 4. Conclusions

*Trichoderma* is one of the most important fungi in agriculture. It has demonstrated many capabilities to be used as biofertilizers as well as biofungicides. It has also shown its sustainability and various mechanisms of providing the crop with nutrients. Moreover, it has various control mechanisms for various plant pathogens, which gives it an advantage when compared to other phytopathogen control mechanisms. It is therefore an option for farmers to use for sustainable cropping and increase in yields and quality of the produce.

## Acknowledgements

Authors would like to acknowledge Dr. Kwasi Sackey Yobo, Bongi Kubheka, Nolitha Skenjana and Sinegugu Shude for their support during the writing of this chapter. We would also like to acknowledge our families for moral support and understanding.

## Conflict of interest

The authors declare no conflict of interest.

## Author details

Bongani Petros Kubheka[1,2]* and Luwam Weldegabir Ziena[2]

1 Dohne Agricultural Development Institute, Stutterheim, South Africa

2 University of KwaZulu Natal, Scottsville, South Africa

*Address all correspondence to: bongakubheka@gmail.com

## References

[1] Suebrasri T, Harada H, Jogloy S, Ekprasert J, Boonlue S. Auxin-producing fungal endophytes promote growth of sunchoke. Rhizosphere. 2020;**16**:100271. DOI: 10.1016/j.rhisph.2020.100271

[2] Macías-Rodríguez L, Contreras-Cornejo HA, Adame-Garnica SG, del-Val E, Larsen J. The interactions of Trichoderma at multiple trophic levels: Inter-kingdom communication. Microbiological Research. 2020;**240**: 126552. DOI: 10.1016/j.micres.2020.126552

[3] Singh B, Boukhris I, Pragya KUMARV, Yadav AN, Farhat-Khemakhem A, et al. Contribution of microbial phytases to the improvement of plant growth and nutrition: A review. Pedosphere. 2020;**30**(3):295-313. DOI: 10.1016/S1002-0160(20)60010-8

[4] Ren K, Hayat S, Qi X, Liu T, Cheng Z. The garlic allelochemical DADS influences cucumber root growth involved in regulating hormone levels and modulating cell cycling. Journal of Plant Physiology. 2018;**230**:51-60. DOI: 10.1016/j.jplph.2018.08.007

[5] Hoyos-carvajal L, Orduz S, Bissett J. Growth stimulation in bean ( Phaseolus vulgaris L .) by Trichoderma. Biological Control. 2009;**51**(3):409-416. DOI: 10.1016/j.biocontrol.2009.07.018

[6] Eslahi N, Kowsari M, Motallebi M, Zamani MR, Moghadasi Z. Influence of recombinant Trichoderma strains on growth of bean (Phaseolus vulgaris L) by increased root colonization and induction of root growth related genes. Scientia Horticulturae. 2020;**261**:108932. DOI: 10.1016/j.scienta.2019.108932

[7] Singh SP, Pandey S, Mishra N, Giri VP, Mahfooz S, Bhattacharya A, et al. Supplementation of Trichoderma improves the alteration of nutrient allocation and transporter genes expression in rice under nutrient deficiencies. Plant Physiology and Biochemistry. 2019;**143**:351-363. DOI: 10.1016/j.plaphy.2019.09.015

[8] Akladious SA, Abbas SM. Application of Trichoderma Harzianum T22 As a biofertilizer potential in maize growth. Journal of Plant Nutrition. 2014;**37**(1):30-49. DOI: 10.1080/01904167.2013.829100

[9] Ghorbanpour M, Omidvari M, Abbaszadeh-Dahaji P, Omidvar R, Kariman K. Mechanisms underlying the protective effects of beneficial fungi against plant diseases. Biological Control. 2018;**117**:147-157. DOI: 10.1016/j.biocontrol.2017.11.006

[10] Gajera HP, Hirpara DG, Savaliya DD, Golakiya BA. Extracellular metabolomics of Trichoderma biocontroller for antifungal action to restrain Rhizoctonia solani Kuhn in cotton. Physiological and Molecular Plant Pathology. 2020;**112**:101547. DOI: 10.1016/j.pmpp.2020.101547

[11] Sornakili A, Thankappan S, Sridharan AP, Nithya P, Uthandi S. Antagonistic fungal endophytes and their metabolite-mediated interactions against phytopathogens in rice. Physiological and Molecular Plant Pathology. 2020;**112**:101525. DOI: 10.1016/j.pmpp.2020.101525

[12] Rajani P, Rajasekaran C, Vasanthakumari MM, Olsson SB, Ravikanth G, Uma SR. Inhibition of plant pathogenic fungi by endophytic Trichoderma spp. through mycoparasitism and volatile organic

compounds. Microbiological Research. 2021;**242**:126595. DOI: 10.1016/j.micres.2020.126595

[13] Kamble MV, Joshi SM, Hadimani S, Jogaiah S. Biopriming with rhizosphere Trichoderma harzianum elicit protection against grapevine downy mildew disease by triggering histopathological and biochemical defense responses. Rhizosphere. 2021;**19**:100398. DOI: 10.1016/j.rhisph.2021.100398

[14] Zhang Y, Zhuang W. Trichoderma brevicrassum strain TC967 with capacities of diminishing cucumber disease caused by Rhizoctonia solani and promoting plant growth. Biological Control. 2020;**142**:104151. DOI: 10.1016/j.biocontrol.2019.104151

[15] Bader AN, Salerno GL, Covacevich F, Consolo VF. Native Trichoderma harzianum strains from Argentina produce indole-3 acetic acid and phosphorus solubilization, promote growth and control wilt disease on tomato (Solanum lycopersicum L.). Journal of King Saud University-Science. 2020;**32**(1):867-873. DOI: 10.1016/j.jksus.2019.04.002

[16] Rafael L, Valadares-inglis MC, Henrique G, Peixoto S, Eliza B, De LG, et al. Volatile organic compounds emitted by Trichoderma azevedoi promote the growth of lettuce plants and delay the symptoms of white mold. Biological Control. 2021;**152**:104447. DOI: 10.1016/j.biocontrol.2020.104447

[17] Juan Z, Ting LIU, Wei-cheng LIU, Dian-peng Z, Dan D, Hui-ling WU, et al. Transcriptomic insights into growth promotion effect of Trichoderma afroharzianum TM2-4 microbial agent on tomato plants. Ournal of integrative. Agriculture. 2021;**20**(5):1266-1276. DOI: 10.1016/S2095-3119(20)63415-3

[18] Ji S, Liu Z, Liu B, Wang Y, Wang J. The effect of Trichoderma biofertilizer on the quality of flowering Chinese cabbage and the soil environment. Scientia Horticulturae. 2020;**262**:109069. DOI: 10.1016/j.scienta.2019.109069

[19] An-le HE, Jia LIU, Xin-hua W, Quan-guo Z, Wei S, Jie C. Soil application of Trichoderma asperellum GDFS1009 granules promotes growth and resistance to fusarium graminearum in maize. Journal of Integrative Agriculture. 2019;**18**(3):599-606. DOI: 10.1016/S2095-3119(18)62089-1

[20] Goh TB, Karamanos RE, Lee J. Effects of phosphorus form on short-term solubility and availability in soils. Communications in Soil Science and Plant Analysis. 2013;**44**(1-4):136-144. DOI: 10.1080/00103624.2013.736162

[21] Mercl F, García-Sánchez M, Kulhánek M, Košnář Z, Száková J, Tlustoš P. Improved phosphorus fertilisation efficiency of wood ash by fungal strains Penicillium sp. PK112 and Trichoderma harzianum OMG08 on acidic soil. Applied Soil Ecology. 2020;**147**:103360. DOI: 10.1016/j.apsoil.2019.09.010

[22] Gasparetti C, Nordlund E, Jänis J, Buchert J, Kruus K. Biochimica et Biophysica Acta extracellular tyrosinase from the fungus Trichoderma reesei shows product inhibition and different inhibition mechanism from the intracellular tyrosinase from Agaricus bisporus. Biochimica et Biophysica Acta (BBA)-Proteins and Proteomics. 2012;**1824**(4):598-607. DOI: 10.1016/j.bbapap.2011.12.012

[23] Yu Z, Wang Z, Zhang Y, Wang Y, Liu Z. Biocontrol and growth-promoting effect of Trichoderma asperellum TaspHu1 isolate from Juglans mandshurica rhizosphere

soil. Microbiological Research. 2021;**242**:126596. DOI: 10.1016/j.micres.2020.126596

[24] Eke P, Wakam LN, Fokou PVT, Ekounda TV, Sahu KP, Kamdem Wankeu TH, et al. Improved nutrient status and fusarium root rot mitigation with an inoculant of two biocontrol fungi in the common bean (Phaseolus vulgaris L.). Rhizosphere. 2019;**12**:100172. DOI: 10.1016/j.rhisph.2019.100172

[25] Yadav RL, Suman A, Prasad SR, Prakash O. Effect of Gluconacetobacter diazotrophicus and Trichoderma viride on soil health , yield and N-economy of sugarcane cultivation under subtropical climatic conditions of India. European Journal of Agronomy. 2009;**30**(4):296-303. DOI: 10.1016/j.eja.2009.01.002

[26] Valentinuzzi F, Cesco S, Tomasi N, Mimmo T. Rhizosphere effect of aluminium exposure on the release of organic acids and genistein from the roots of Lupinus albus L . Plants. Rhizosphere. 2016;**1**:29-32. DOI: 10.1016/j.rhisph.2016.07.002

[27] Swain H, Adak T, Mukherjee AK, Mukherjee PK, Bhattacharyya P, Behera S, et al. Novel Trichoderma strains. Isolated from tree barks as potential biocontrol agents and biofertilizers for direct seeded rice. Microbiological Research. 2018;**214**:83-90. DOI: 10.1016/j.micres.2018.05.015

[28] Sani MNH, Hasan M, Uddain J, Subramaniam S. Impact of application of Trichoderma and biochar on growth, productivity and nutritional quality of tomato under reduced N-P-K fertilization. Annals of Agricultural Science. 2020;**65**(1):107-115. DOI: 10.1016/j.aoas.2020.06.003

[29] Khan Y, Haque M, Molla AH, Rahman M. Antioxidant compounds and minerals in tomatoes by Trichoderma - enriched biofertilizer and their relationship with the soil environments. Journal of Integrative Agriculture. 2017;**16**(3):691-703. DOI: 10.1016/S2095-3119(16)61350-3

[30] De SA, García-lópez AM, Manuel J, Avilés M, Delgado A. Soil Biology & Biochemistry Effect of Trichoderma asperellum strain T34 and glucose addition on iron nutrition in cucumber grown on calcareous soils. Soil Biology and Biochemistry. 2013;**57**:598-605. DOI: 10.1016/j.soilbio.2012.06.020

[31] Zin NA, Badaluddin NA. Biological functions of Trichoderma spp. for agriculture applications. Annals of Agricultural Science. 2020;**65**(2):168-178. DOI: 10.1016/j.aoas.2020.09.003

[32] Rokni N, Shams Alizadeh H, Bazgir E, Darvishnia M, Mirzaei NH. The tripartite consortium of Serendipita indica, Trichoderma simmonsii, and bell pepper (Capsicum annum). Biological Control. 2021;**158**:104608. DOI: 10.1016/j.biocontrol.2021.104608

[33] Tripathi P, Singh PC, Mishra A, Tripathi RD, Nautiyal CS. Ecotoxicology and environmental safety Trichoderma inoculation augments grain amino acids and mineral nutrients by modulating arsenic speciation and accumulation in chickpea ( Cicer arietinum L .). Ecotoxicology and Environmental Safety. 2015;**117**:72-80. DOI: 10.1016/j.ecoenv.2014.10.027

[34] Yin J, Yuan L, Sui Z, Huang J. Mobilization of organic nitrogen and phosphorus and reduction of synthetic fertilizer usage by Ceriporia lacerata HG2011 in pepper cultivation. Scientia Horticulturae. 2021;**293**:110721. DOI: 10.1016/j.scienta.2021.110721

[35] Pellegrini A, Prodorutti D, Pertot I. Use of bark mulch pre-inoculated with

Trichoderma atroviride to control Armillaria root rot. Crop Protection. 2014;**64**:104-109. DOI: 10.1016/j.cropro.2014.06.007

[36] Rocha FYO, de Oliveira CM, da Silva PRA, de Melo LHV, do Carmo MGF, Baldani JI. Taxonomical and functional characterization of bacillus strains isolated from tomato plants and their biocontrol activity against races 1, 2 and 3 of fusarium oxysporum f. sp. Lycopersici. Applied Soil Ecology. 2017;**120**:8-19. DOI: 10.1016/j.apsoil.2017.07.025

[37] De Corato U, Pane C, Bruno GL, Cancellara FA, Zaccardelli M. Co-products from a biofuel production chain in crop disease management: A review. Crop Protection. 2015;**68**:12-26. DOI: 10.1016/j.cropro.2014.10.025

[38] Adnan M, Islam W, Shabbir A, Khan KA, Ghramh HA, Huang Z, et al. Plant defense against fungal pathogens by antagonistic fungi with Trichoderma in focus. Microbial Pathogenesis. 2019;**129**:7-18. DOI: 10.1016/j.micpath.2019.01.042

[39] Jaber LR, Alananbeh KM. Fungal entomopathogens as endophytes reduce several species of fusarium causing crown and root rot in sweet pepper (Capsicum annuum L.). Biological Control. 2018;**126**:117-126. DOI: 10.1016/j.biocontrol.2018.08.007

[40] Adnan M, Islam W, Shabbir A, Khan KA, Ghramh HA, Huang Z, et al. Plant defense against fungal pathogens by antagonistic fungi with Trichoderma in focus. Microbial Pathogenesis. 2019;**129**:7-18. DOI: 10.1016/j.micpath.2019.01.042

[41] Chen S, Yan Y, Wang Y, Wu M, Mao Q, Chen Y, et al. Trichoderma asperellum reduces phoxim residue in roots by promoting plant detoxification potential in Solanum lycopersicum L. Environmental Pollution. 2020;**259**:113893. DOI: 10.1016/j.envpol.2019.113893

[42] Ojwach J, Kumar A, Mutanda T, Mukaratirwa S. Fructosyltransferase and inulinase production by indigenous coprophilous fungi for the biocatalytic conversion of sucrose and inulin into oligosaccharides. Biocatalysis and Agricultural Biotechnology. 2020;**30**:101867. DOI: 10.1016/j.bcab.2020.101867

[43] Chao W, Wen-ying Z. Evaluating effective Trichoderma isolates for biocontrol of Rhizoctonia solani causing root rot of Vigna unguiculata. Journal of Integrative Agriculture. 2019;**18**(9):2072-2079. DOI: 10.1016/S2095-3119(19)62593-1

[44] De la Cruz-Quiroz R, Roussos S, Rodríguez-Herrera R, Hernandez-Castillo D, Aguilar CN. Growth inhibition of Colletotrichum gloeosporioides and Phytophthora capsici by native Mexican Trichoderma strains. Karbala International Journal of Modern Science. 2018;**4**(2):237-243. DOI: 10.1016/j.kijoms.2018.03.002

[45] Marzano M, Gallo A, Altomare C. Improvement of biocontrol efficacy of Trichoderma harzianum vs . Fusarium oxysporum f . Sp . Lycopersici through UV-induced tolerance to fusaric acid. Biological Control. 2013;**67**(3):397-408. DOI: 10.1016/j.biocontrol.2013.09.008

[46] Alizadeh H, Behboudi K, Ahmadzadeh M, Javan-Nikkhah M, Zamioudis C, Pieterse CMJ, et al. Induced systemic resistance in cucumber and Arabidopsis thaliana by the combination of Trichoderma harzianum Tr6 and pseudomonas sp. Ps14. Biological Control. 2013;**65**(1):14-23. DOI: 10.1016/j.biocontrol.2013.01.009

[47] Chilosi G, Aleandri MP, Luccioli E, Stazi SR, Marabottini R, Morales-Rodríguez C, et al. Suppression of soil-borne plant pathogens in growing media amended with espresso spent coffee grounds as a carrier of Trichoderma spp. Scientia Horticulturae. 2020;**259**:108666. DOI: 10.1016/j.scienta.2019.108666

[48] Aleandri MP, Chilosi G, Bruni N, Tomassini A, Vettraino AM, Vannini A. Use of nursery potting mixes amended with local Trichoderma strains with multiple complementary mechanisms to control soil-borne diseases. Crop Protection. 2015;**67**:269-278. DOI: 10.1016/j.cropro.2014.10.023

[49] Victoria Arellano AD, Silva GM da, Guatimosim E, Dorneles K da R, Moreira LG, Dallagnol LJ. Seeds coated with Trichoderma atroviride and soil amended with silicon improve the resistance of Lolium multiflorum against Pyricularia oryzae. Biological Control 2021; 154:104499. DOI: 10.1016/j.biocontrol.2020.104499

[50] Bansal R, Pachauri S, Gururajaiah D, Sherkhane PD, Khan Z, Gupta S, et al. Dual role of a dedicated GAPDH in the biosynthesis of volatile and non-volatile metabolites- novel insights into the regulation of secondary metabolism in Trichoderma virens. Microbiological Research. 2021;**253**:126862. DOI: 10.1016/j.micres.2021.126862

[51] Mukhopadhyay R, Kumar D. Trichoderma: A beneficial antifungal agent and insights into its mechanism of biocontrol potential. Egyptian Journal of Biological Pest Control. 2020;**30**(1):1-8. DOI: 10.1186/s41938-020-00333-x

[52] Govarthanan M, Mythili R, Selvankumar T, Kamala-Kannan S, Kim H. Myco-phytoremediation of arsenic- and lead-contaminated soils by Helianthus annuus and wood rot fungi, Trichoderma sp. isolated from decayed wood. Ecotoxicology and Environmental Safety. 2018;**151**:279-284. DOI: 10.1016/j.ecoenv.2018.01.020

[53] da Delabona PS, Lima DJ, Codima CA, Ramoni J, Gelain L, de Melo VS, et al. Replacement of the carbon catabolite regulator (cre1) and fed-batch cultivation as strategies to enhance cellulase production in Trichoderma harzianum. Bioresource Technology Reports. 2021;**13**:1-7. DOI: 10.1016/j.biteb.2021.100634

[54] De Palma M, D'Agostino N, Proietti S, Bertini L, Lorito M, Ruocco M, et al. Suppression subtractive hybridization analysis provides new insights into the tomato (Solanum lycopersicum L.) response to the plant probiotic microorganism Trichoderma longibrachiatum MK1. Journal of Plant Physiology. 2016;**190**:79-94. DOI: 10.1016/j.jplph.2015.11.005

[55] Mironenka J, Różalska S, Soboń A, Bernat P. Lipids, proteins and extracellular metabolites of Trichoderma harzianum modifications caused by 2,4-dichlorophenoxyacetic acid as a plant growth stimulator. Ecotoxicology and Environmental Safety. 2020;**194**:110383. DOI: 10.1016/j.ecoenv.2020.110383

[56] Almeida DA, Horta MAC, Ferreira Filho JA, Murad NF, de Souza AP. The synergistic actions of hydrolytic genes reveal the mechanism of Trichoderma harzianum for cellulose degradation. Journal of Biotechnology. 2021;**334**:1-10. DOI: 10.1016/j.jbiotec.2021.05.001

[57] Sharma R, Magotra A, Manhas RS, Chaubey A. Antagonistic potential of a psychrotrophic fungus: Trichoderma velutinum ACR-P1. Biological Control. 2017;**115**:12-17. DOI: 10.1016/j.biocontrol.2017.08.024

[58] Näsi M, Piironen J, Partanen K. Efficacy of Trichoderma reesei phytase and acid phosphatase activity ratios in phytate phosphorus degradation in vitro and in pigs fed maize-soybean meal or barley-soybean meal diets. Animal Feed Science and Technology. 1999;**77**(1-2):125-137. DOI: 10.1016/S0377-8401(98)00241-7

[59] Poveda J, Baptista P. Filamentous fungi as biocontrol agents in olive (Olea europaea L.) diseases: Mycorrhizal and endophytic fungi. Crop Protection. 2021;**146**:105672. DOI: 10.1016/j.cropro.2021.105672

[60] Usman Ghazanfar M, Raza M, Raza W, Qamar MI. Plant Protection Trichoderma as potential biocontrol agent, its exploitation in agriculture: A Review [Internet]. 2018. Available from: https://www.academia.edu/38806681/Trichoderma_as_potential_biocontrol_agent_its_exploitation_in_agriculture_A_review [Accessed: 2021-11-25]

[61] Mukesh S, Vipul K, Mohammad S, Sonika P, ey, Anuradha S. Trichoderma- a potential and effective bio fungicide and alternative source against notable phytopathogens: A review. African Journal of Agricultural Research. 2016;**11**(5):310-316. DOI: 10.5897/AJAR2015.9568

[62] Devi KA, Pandey P, Sharma GD. Plant growth-promoting endophyte Serratia marcescens AL2-16 enhances the growth of Achyranthes aspera L., a medicinal plant. HAYATI. Journal of Biosciences. 2016;**23**(4):173-180. DOI: 10.1016/j.hjb.2016.12.006

[63] Deng JJ, Shi D, Hua MH, Wei LZ, Liang S, Ke Y, et al. Heterologous expression and characterization of an antifungal chitinase (Chit46) from Trichoderma harzianum GIM 3.442 and its application in colloidal chitin conversion. International Journal of Biological Macromolecules. 2019;**134**:113-121. DOI: 10.1016/j.ijbiomac.2019.04.177

[64] Alizadeh H, Behboudi K, Ahmadzadeh M, Javan-nikkhah M, Zamioudis C, Pieterse CMJ, et al. Induced systemic resistance in cucumber and Arabidopsis thaliana by the combination of Trichoderma harzianum Tr6 and pseudomonas sp . Ps14. Biological Control. 2013;**65**(1):14-23. DOI: 10.1016/j.biocontrol.2013.01.009

[65] Berendsen RL, Vismans G, Yu K, Song Y, de Jonge R, Burgman WP, et al. Disease-induced assemblage of a plant-beneficial bacterial consortium. The ISME Journal. 2018;**12**(6):1496-1507. DOI: 10.1038/s41396-018-0093-1

[66] Bastias DA, Martínez-Ghersa MA, Ballaré CL, Gundel PE. Epichloë fungal endophytes and plant Defenses: Not just alkaloids. Trends in Plant Science. 2017;**22**(11):939-948. DOI: 10.1016/j.tplants.2017.08.005

[67] Martínez-medina A, Roldán A, Albacete A, Pascual JA. Phytochemistry the interaction with arbuscular mycorrhizal fungi or Trichoderma harzianum alters the shoot hormonal profile in melon plants. Phytochemistry. 2011;**72**:223-229. DOI: 10.1016/j.phytochem.2010.11.008

[68] Wang Y, Yuan J, Li S, Hui L, Li Y, Chen K, et al. Comparative analysis of carbon and nitrogen metabolism, antioxidant indexes, polysaccharides and lobetyolin changes of different tissues from Codonopsis pilosula co-inoculated with Trichoderma. Journal of Plant Physiology. 2021;**267**:153546. DOI: 10.1016/j.jplph.2021.153546

[69] Morán-Diez ME, Tranque E, Bettiol W, Monte E, Hermosa R. Differential response

of tomato plants to the application of three trichoderma species when evaluating the control of pseudomonas syringae populations. Plants. 2020;**9**(5):626. DOI: 10.3390/plants9050626

[70] Hermosa R, Belén Rubio M, Cardoza RE, Nicolás C, Monte E, Gutiérrez S. The contribution of Trichoderma to balancing the costs of plant growth and defense. International Microbiology. 2013;**16**(2):69-80. DOI: 10.2436/20.1501.01.181

[71] Guo R, Ji S, Wang Z, Zhang H, Wang Y, Liu Z. Trichoderma asperellum xylanases promote growth and induce resistance in poplar. Microbiological Research. 2021;**248**:126767. DOI: 10.1016/j.micres.2021.126767

[72] Lucini L, Colla G, Miras Moreno MB, Bernardo L, Cardarelli M, Terzi V, et al. Inoculation of Rhizoglomus irregulare or Trichoderma atroviride differentially modulates metabolite profiling of wheat root exudates. Phytochemistry. 2019;**157**:158-167. DOI: 10.1016/j.phytochem.2018.10.033

[73] Aamir M, Kashyap SP, Zehra A, Dubey MK, Singh VK, Ansari WA, et al. Trichoderma erinaceum bio-priming modulates the WRKYs Defense programming in tomato against the fusarium oxysporum f. sp. lycopersici (Fol) challenged condition. Frontiers in Plant Science. 2019;**10**:911. DOI: 10.3389/fpls.2019.00911

[74] Mei LI, Hua LIAN, Su XL, Tian Y, Huang WK, Mei J, et al. The effects of Trichoderma on preventing cucumber fusarium wilt and regulating cucumber physiology. Journal of Integrative Agriculture. 2019;**18**(3):607-617. DOI: 10.1016/S2095-3119(18)62057-X

[75] Shi M, Chen L, Wang XW, Zhang T, Zhao PB, Song XY, et al. Antimicrobial peptaibols from Trichoderma pseudokoningii induce programmed cell death in plant fungal pathogens. Microbiology. 2012;**158**(1):166-175. DOI: 10.1099/mic.0.052670-0

[76] Viterbo A, Haran S, Friesem D, Ramot O, Chet I. Antifungal activity of a novel endochitinase gene (chit36) from Trichoderma harzianum Rifai TM. FEMS Microbiology Letters. 2001;**200**(2):169-174. DOI: 10.1111/j.1574-6968.2001.tb10710.x

[77] Zhang F, Chen C, Zhang F, Gao L, Liu J, Chen L, et al. Trichoderma harzianum containing 1-aminocyclopropane-1-carboxylate deaminase and chitinase improved growth and diminished adverse effect caused by fusarium oxysporum in soybean. Journal of Plant Physiology. 2017;**210**:84-94. DOI: 10.1016/j.jplph.2016.10.012

[78] Guzmán-Guzmán P, Alemán-Duarte MI, Delaye L, Herrera-Estrella A, Olmedo-Monfil V. Identification of effector-like proteins in Trichoderma spp. and role of a hydrophobin in the plant-fungus interaction and mycoparasitism. BMC Genetics. 2017;**18**(1):1-20. DOI: 10.1186/s12863-017-0481-y

[79] Ortega-García JG, Montes-Belmont R, Rodríguez-Monroy M, Ramírez-Trujillo JA, Suárez-Rodríguez R, Sepúlveda-Jiménez G. Effect of Trichoderma asperellum applications and mineral fertilization on growth promotion and the content of phenolic compounds and flavonoids in onions. Scientia Horticulturae. 2015;**195**:8-16. DOI: 10.1016/j.scienta.2015.08.027

[80] Woo SL, Ruocco M, Vinale F, Nigro M, Marra R, Lombardi N, et al. Trichoderma-based products and their widespread use in agriculture. The Open Mycology Journal. 2014;**8**(1):71-126. DOI: 10.2174/1874437001408010071

Section 2

# Crop Diseases Management

Chapter 3

# *Trchoderma* Spp.: Their Impact in Crops Diseases Management

*Amar Bahadur and Pranab Dutta*

## Abstract

*Trichoderma* species, a cosmopolitan fungi, present in all types of soil, manure, and decaying plant tissues that can degrade domestic waste relatively quickly without emitting bad odors. *Trichoderma* is recognized worldwide as potential fungal bio-control agents for the management of various foliar and soil-borne plant pathogens, highly compatible with sustainable agriculture and play major role as a component of integrated pest management. Bio-control agents are an antagonism and eco-friendly approach for managing plant diseases. *Trichoderma* as bioagent area effective not only against soil-borne plant pathogens, but also against nematodes without any adverse effect on beneficial microbes. *Trichoderma* is capable of growth promotions in crops. There are two major mass production methods of *Trichoderma* spp. *viz*., solid state fermentation and liquid state fermentation. In solid, fungus is grown on various cereal grains, agricultural wastes, and byproducts, and these products are used mainly for direct soil application to suppress the soil-borne inoculums. In a liquid state, *Trichoderma* is grown on media such as molasses and yeast in deep tanks and fermentation can be made into different formulations such as dusts, granules, pellets, wettable powders. As seed-treating agents or bio-priming agents, *Trichoderma* formulations can be successfully used against several soil-borne diseases caused by *Pythium, Phytophthora, Rhizoctonia, Fusarium and Sclerotium, spp*. in several crops.

**Keywords:** *Trchoderma* spp. formulation, multiplication, mechanisms, management

## 1. Introduction

The genus *Trichoderma* strains are versatile, highly competent, root colonizers, cosmopolitan in nature, fast growing in culture; produce numerous green spores and chlamydospores as used for eco-friendly disease management which is important in organic agriculture. *Trichoderma* species have been used as a biological control, biofertilizers, source of enzymes and protein producers. *Trichoderma* increase the fertility of soils and improved plant growth beyond disease control [1, 2]. *Trichoderma* strains colonization on root and enhances root growth, root area, root length, increases in dry weight, shoot length and leaf area [3]. Plant growth promotion is due to the production of plant hormones and the uptake of nutrients by the plant [4]. They promote root growth, nutrient availability and release plant growth regulators [5]. Application in plants and can prevent the infection of diseases through induced resistance, competition for nutrients and space, antibiosis, hyperparasitism. Induced resistance may be local or systemic. *Trichoderma* species are cosmopolitan fungi, frequently present in all types of soil, manure and decaying plant tissues [6]. *Trichoderma* species are a well known bio-control agent, utilize chitinolytic enzymes to disintegrate and degrade the pathogen's cell walls and colonize on the root, soil and foliar environments suppressing phytopathogens [7–9]. *Trichoderma* spp. is highly interactive in root, soil and foliar in the environment, parasitize other fungi [10]. *Trichoderma* was first described in 1794 and its perfect stage (*Hypocrea*). Morphological characters and an online identification tool are available for identification of species within the genus *Trichoderma* and recognized from long back as biological agents, control of plant disease and also their ability to increase root growth and development, crop productivity, resistance to abiotic stresses, and uptake and use of nutrients. Application of *Trichoderma harzianum* to plants resulted in improved seed germination, increased plant size, and augment of leaf area and weight [11]. *Trichoderma* spp. is well documented and effective biological control agent of soil-borne diseases by secreting several cell wall degrading enzymes, antibiotics [12, 13]. They have produced extracellular proteins and fungi toxic substances for understanding the role in antagonistic as playing in biological interactions [14]. *Trichoderma reesei* and *Trichoderma harzianum* are capable of producing proteinase, mananase, laminarinase and chitinase that the nature of antagonism by mycoparasitism [15]. *Trichoderma*, a soil-borne mycoparasitic fungus has been shown effective against many.soil borne phytopathogens [16–19]. *Trichoderma viride* and *Trichoderma harzianum* found to highly antagonistic against *Sclerotium rolfsii* and management of diseases in vegetables and legumes [20–29]. *Trichoderma harzianum* is an antagonist probably their ability to fight for cellulose in the mucilage layer at the root surface [30]. *Trichoderma* spp. have been reported to control soil-borne plant pathogens viz., *Rhizoctonia solani* Khun., *Sclerotium rolfsii* (Sacc.) Curzi., *Pythium* and *Fusarium* spp. [31–36] and *Botrytis* rot of fruits on grape. *Trichoderma harzianum*, widely tested as potential biological control agents many soil-borne plant pathogens [10, 37–39]. *Trichoderma hamatum* produced inhibitory volatiles compounds that reduced the gray mold of snap bean pods and blossom [40, 41]. *Trichoderma harzianum* T39 less effective in cucumber fruit and stem gray mold under wet and below 20°C compare to elevated temperatures [42]. *Trichoderma harzianum* inoculated in root increased peroxidase and chitinase activities in leaves of cucumber seedlings [43]. *Trichoderma harzianum* T39 effectively controlled *Botrytis* diseases, white mold (*Sclerotinia sclerotiorum*), leaf mold (*Cladosporium fulvum*) and powdery mildew (*Sphaerotheca fusca*) [42, 44–47]. *Trichoderma* bio-control agents are used against fungal phyto-pathogens such as *Phythium, Phytophthora*, Macrophomina, *Aspergillus*, *Rhizoctonia* and *Fusarium* through the mechanism of mycoparasitism, antibiotics and competition for food and space [5, 48]. Induced resistance is an important mode of bio-control in vegetative tissues [49, 50]. *T. harzianum* induced systemic resistance in the plant against fungal and bacterial pathogens [8, 9]. Induced systemic resistance caused by various micro-organisms and protects plants against the soil or foliar pathogens [51]. *Trichoderma*, *Gliocladium* and *Pythium* spp. are known as mycoparasites [46]. *Pythium* spp. is non specific mycoparasites and interact with many soil borne fungi [52]. *Trichoderma virens* produces two major antifungal antibiotics- gliotoxin (toxic to Rhizoctonia *solani* and *Pythium ultimum*) and gliovirin (toxic to *Pythium* spp.) [53]. *Trichoderma virens*, produce the antibiotics gliovirin and gliotoxin as mycoparasite. *Trichoderma harzianum* T39 is competition for nutrients and interference with the production of pectolytic enzymes against the pathogen, and also prevents the penetration of the host tissue and is shown to induce resistance [54–56]. *Trichoderma harzianum* T39 produces protease on leaves against *Botrytis cinerea* disease development [57]. Secretion of proteolytic enzymes that deactivate pathogenicity related hydrolytic enzymes of pathogenic fungi [58]. *Trichoderma viride* is involved in bio-control management *Sclerotium rolfsii* through the proteolytic activity as reported [59]. *Trichoderma* species can degrade domestic waste quickly without emitting bad odors [60]. The genus Trichoderma has the potential to control plant-parasitic nematodes [7, 61]. *Trichoderma harzianum* is associated with a reduction in nematode population by parasitizing and killing in the rhizosphere [62]. *T. harzianum* has a rich source of chitinolytic enzymes which might degrade the eggshell during parasitism of eggs and juveniles [39, 62]. *Trichoderma* promotes crop productivity, resistance to abiotic stresses and uptake of nutrients [63]. *Trichoderma* colonization in the roots and soil helps insolubilization of minerals viz.; rock phosphate, Fe, Mn, Cu and Zn and also enhances N-used efficiency [1]. *Trichoderma* based commercial products are manufactured and marketed worldwide for the management of plant diseases [10, 17].

## 2. Characters and isolation techniques

*Trichoderma* is a good bio-control agent as well as a fertility promoter. *Trichoderma* fungi can produce antibiotics, enzymes that antagonize plant pathogens and hormones that regulate root architecture and promote plant growth. *Trichoderma* protects a wide range of foliar pathogens. *Trichoderma* strains have numerous mechanisms for attacking other fungi and enhancing plant and root growth [1]. *Trichoderma* is colonizing in the rhizosphere and resulted in the increased root, aerial systems and crop yields [64, 65]. *Trichoderma* has a strong capacity to mobilize and take up soil nutrients and making it more efficient and competitive. Competition for nutrients is the major mechanism used by *Trichoderma harzianum* to control *Fusarium oxysporum* f. sp. *melonis. Trichoderma* spp. produces three kinds of propagules; hyphae, chlamydospores, and conidia [18]. The main propagules of *Trichoderma* spp. is hyphae and during the drying process losing viability. Chlamydospores and conidia have been used as the active ingredients of *Trichoderma* spp. based production [66–68]. *Trichoderma* fungal conidiophores are highly branched, not verticilate, 1-celled, ovoid, borne in small terminal clusters, loosely or compactly tufted, and often formed in distinct concentric rings (**Figure 1**). The conidiophore branches with paired and assume a pyramidal feature, chlamydospores produced by all species. *Trichoderma* strains produce only asexual spores, the sexual stage of *Trichoderma* belong to the ascomycete genus *Hypocrea* [69]. *Trichoderma* is one of the common fungal bio-control agents used worldwide for the management of various foliar and soil borne plant pathogens. *Trichoderma* fungi are present in most of all types of soils can be isolated from forest and agriculture soils and wood. Several species are beneficial in agriculture. The fungus grows a range between 25 and 30°C of optimal. The most suitable culture media for its cultivation are cornmeal dextrose agar whereas colonies appear transparent, and on potato dextrose agar the colonies appear initially white and then green (**Figure 2**). *Trichoderma* can be produced in a liquid or solid fermentation medium.

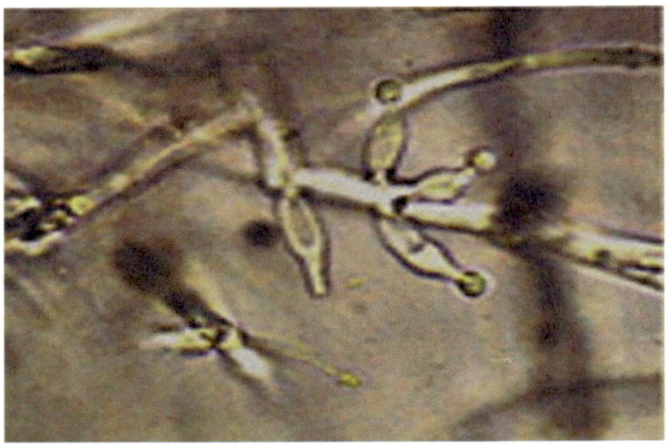

**Figure 1.**
*Phialid showing production of conidia in highly branched conidiophore (photo credit: Dr. Pranab Dutta & Lipa deb, CPGSAS, CAU (Imphal), Umiam, Meghalaya.*

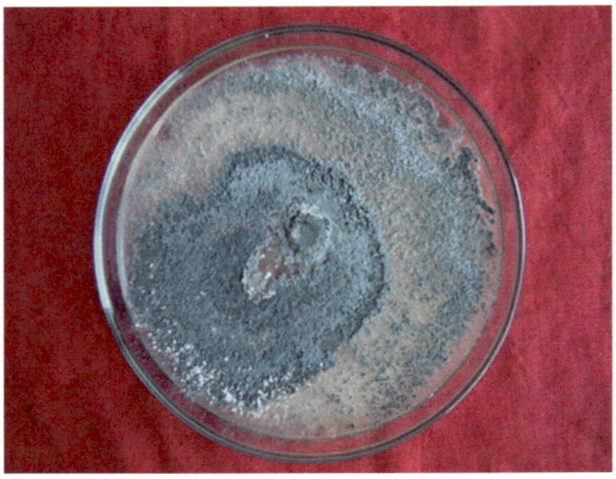

**Figure 2.**
*Pure culture of Trichoderma harzianum in PDA plate (photo credit: Dr. Pranab Dutta, CPGSAS, CAU (Imphal), Umiam, Meghalaya.*

## 3. Mass multiplication

The mass production of *Trchoderma* for field application and commercial use. There are two methods of production as (i) Solid-state fermentation (ii) Liquid-state fermentation.

### 3.1 Solid-state fermentation

It is a common method for mass production of *Trichoderma*. The *Trichoderma* spp. have been widely grown on solid substrates like sorghum grain, wheat straw, wheat bran, spent tea leaf waste, coffee husk, rice husk, banana leaves, sawdust etc. and their mass multiplication [70–72]. The commonly used for mass culturing of *Trichoderma* spp. on the solid substrate as sorghum grain [73], wheat bran-saw dust [74] and other agro-based waste products. Cereal grains like, sorghum, millets, ragi are used as substrates [75]. The grains are moistened, sterilized and inoculated with *Trichoderma* and incubated for 10–15 days. The dark green spore coating on the grains of *Trichoderma* produces. These grains can be powdered and used as a seed treatment or grains may use as it is for enhancing FYM for soil application.

Wastage substrate potato peel, brinjal, spinach, sugarcane, banana, papaya, guava, tea leaves and pea husk used for the multiplication of *Trichoderma harzianum* and *Trichoderma viride. Trichoderma harzianum* multiplied on presoaked and autoclaved Jhangora seeds for 12 days at 28°C, air dried, ground and passed through 50 and 80 mesh sieves simultaneously to obtain the powder of spores [76]. The commercial formulation was prepared by diluting this powder with the talcum powder containing 1% carboxymethyl-cellulose to get the desired concentration of biocontrol agent.

### 3.2 Liquid fermentation

*Trichoderma* is grown in a liquid fermentation system on media in stationary/shaker/fermentor and used for field application. The production of *Trichoderma* in liquid state fermentation includes molasses and brewer's yeast [77], and Jaggery-soy medium [32, 33]. Viable propagules of *Trichoderma harzianum and Trichoderma viride* can be obtained within 96 h of fermentation in a fermentor with aeration, agitation, temperature controls [78]. Maximum biomass of Trichoderma spp. in short-time by using the appropriate medium in a fermentor with aeration, agitation, temperature, pH with antifoam controls than in shake-flask cultures and more suitable for industrial production. The liquid fermentation can be separated of biomass with medium and incorporated into dust, granules, pellets, wettable powders or emulsifiable liquids. The carrier is inert as the food base of *Trichoderma* spp. can be formulated as pellets [79], dust and powders [40, 41] and fluid drill gels [21]. Molasses yeast medium is used for mother culture; it's prepared by adding of molasses 30 g, yeast 5 g and distiller water 1000 ml. The medium hand out into conical flasks and sterilized at 15 lb. pressure for 15 minutes in an autoclave. After the cooled medium is inoculated with 10 days old fungal disc of *Trichoderma viride* and then incubated for 10 days for fungal growth, serves as a mother culture. The mother culture was added to the fermentor at the rate of 1.5 lit/50 lit of the medium and incubated at room temperature for 10 days. Then the incubated fungal culture is used for commercial formulation preparation using talc powder.

## 4. Formulations and application

The formulation depends on the type of application, its combination of active ingredients, such as fungal spores with the inert material as diluents of the desirable form. The formulation developed through standard air dried mats and mixed with the carrier contain $10^8$–$10^9$ propagules per gram [80]. *Trichoderma* is grown in the liquid medium is mixed with talc powder in the ratio of 1: 2 and dried to 8% moisture under shade. Bio-control formulation, distribution and execlution of microbial antagonists [81]. Talc based formulations of *Trichoderma viride* developed for seed treatment of pulse and rice crops [82]. The commercial formulations of *Trichoderma* spp. based on carriers are available for controlling plant diseases [83]. *Trichoderma* formulation prepared based on coffee husk which is a waste from the coffee curing industry [84]. *Trichoderma* was formulated on press mud to farmers and value-added organic manure by sugar factory [75]. Oil-based formulations are suitable for foliar sprays under dry weather conditions with prolonged shelf life. *T. harzianum* is an emulsion based formulation with a shelf life of 8 months use for the control of post -harvest decay of apple caused by *Botrytis cinerea* [37]. The application of *Trichoderma* is very important for successful diseases management. Das et al. [24] tested three different media amended with and without osmoticant (Mnaitol) viz., potato dextrose broth (PDB), modified Richard's broth (MRB) and Czapek dox broth (CDB) were tested for biomass production of *Trichoderma harzianum*. Osmoticant amended MRB was found best for production of maximum sporulation, cfu and dry weight of biomass of the antagonist.

The common methods are seed treatment, seed bio-priming, seedling root dip, and soil application and wound dressing.

### 4.1 Seed treatment

Seed coating with dry powder of *Trichoderma* just before sowing is an effective method of antagonist for the management of soil-borne diseases. Seeds coated with a commercial dry powder of *Trichoderma* just before sowing at 3 to 10 g/kg seed based on seed size [85]. Seed treatment with talc-based and wheat bran based formulations use at 4 g/kg of seed have been recommended [74]. *Trichoderma* germinates on treated seed surface as they are sown in the soil; on germinating propagules colonize the seedlings roots and rhizosphere [86]. *Trichoderma harzianum, Trichoderma virens* and *Trichoderma viride* were found to be effective seed protectants against *Pythium* spp. and *Rhizoctonia solani* [87]. Seed treatments are effective against the sheath blight of rice [25], loose smut of wheat [88]. Das et al. [24] reported that seed treatment with osmoticant amended talc based bioformulation of *Trichoderma harzianum* was best in reduction per cent disease index of soybean rot caused by *Rhizoctonia solani*. Seed treatment with *Trichoderma harzianum* was found has most effective in improving seed germination (18.43%), reducing 90.46% infection by *Rhizoctonia solani* in soybean and increasing yield (69.51%) over control plot [89].

### 4.2 Seed bio-priming

Treated seeds with *Trichoderma* incubate until radical emergence is referred to as bio-primming. This technique is a simple coating of seeds and results in rapid and uniform seedling emergence and also reduces the amount of bio-control agents. *Trichoderma* conidia germinate on the seed surface and form a layer around seeds. Such seeds tolerate adverse conditions of the soil better than the non-primed seeds. Seed bio-priming was successfully used in tomato, brinjal, soybean and chickpea [90]. It results in rapid and uniform seedling emergence and reduces the amount of bio-control agents [91].

### 4.3 Seedling root dip treatment

It is suitable for transplanting rice and vegetable crops. The seedlings can be treated with the spore suspension by mixing 10 g of *Trichoderma* powder with 100 g of well rotten FYM per liter of water and dipping roots for 10 minutes before transplanting/drenching in nursery beds. This method is generally used for vegetable crops, rice where transplanting is practiced [92]. Root dipping in antagonist's suspension reduces disease severity and enhances seedling growth in rice, tomato, brinjal, chili and capsicum as reported [92]. The reduction of sheath blight disease of rice by root dipping in spore suspension of seedlings before transplantation [93]. Root dipping of tomato seedlings reduces the severity of root- knot nematode (*Meloidogyne incognita*).

### 4.4 Soil treatment

Trichoderma is capable of colonizing on farmyard manure (FYM) and then applied to the soil is the most effective method for the management of soil-borne diseases. The application of bio-control agents to the soil before or at the time of planting for control of soil- borne fungal pathogens [94]. Soil application of *Trichoderma viride* alone and in combination reduced red rot caused by *Colletotrichum falcatum* [95] and seedling blight, stem rot, color rot and root rot disease of Jute [14, 96]. Soil treatment with 5 Kg of *Trichoderma* powder per hector mixed after turning of sun hemp/dhaincha into the soil for green manuring or 1 kg of *Trichoderma* formulation in 100 kg of FYM. Some species of *Trichoderma* are reported to cause green mold disease of mushrooms [97]. Soil application (2%) of *Trichoderma harzianum* enriched enriched farm yard manure showed excellent result in reduction of stem rot disease incidence caused by *Rhizoctonia solani* and collarot of disease of tomato caused by *Sclerotium rolfsii* with increased seed germination (%), plant growth parameters and yield of the crop [38, 98].

### 4.5 Aerial spraying/wound dressing

*Trichoderma* can reduce the severity of diseases under field conditions. *Trichoderma* has been successfully applied to the aerial plant parts and on wounds of shrubs and trees [18]. Suspension of *Trichoderma* has been successfully applied to the aerial plant parts infected with *Alternaria* leaf spot of *Vicia faba* [99]. *Trichoderma harzianum* and *Trichoderma* virens talc-based formulations use for foliar sprays that reduce disease incidence of sheath blight of rice [100, 101]. Affects the efficacy and survival of antagonist in phylloplane [102]. The dosage and application have to be standardized based on the crop value. Foliar spraying of consortial formulation of *Trichoderma harzianum* + two entomopathogeic fungi along with seed treatment, seedling root treatment showed significant reduction of three important diseases of brinjal viz., *Phomopsis* leaf blight and fruit rot, *Alternaria* leaf spot, and *Fusarium* wilt [103].

## 5. Viability in the storage and field

One of the critical problems in the commercialization of bio-agents is the loss of viability of the propagules over time. The shelf life of the bio-control product is dependent on the storage temperature and carriers as used in the formulation of bio control agents. The shelf life of bio- control agent plays a significant role in successful marketing. *Trichoderma* spp. are multiplied on bio-degradable substrates for long shelf-life and is also beneficial for field application. Bio-control agents are a biomass product, maintaining their viability at the end of the course [104]. Talc based *Trichoderma virens* conidia keep 82% viability at 5°C in refrigerator after 6 months, while at room temperature was observed for 3 months [105]. The viable propagule of *Trichoderma* in talc formulation was reduced by 50% after 120 days of storage [77]. Increasing shelf life of talc formulations of *Trichoderma* using various ingredients (chitin and glycerol) in production medium fermentation was carried out the shelf of talc formulation of *Trichoderma* up to 1 year [106, 107]. *Trichoderma* on coffee husk has a shelf life of more than 18 months. Talc, peat, lignite and kaolin based formulations of *Trichoderma*, have a shelf life of 3–4 months. In the storage polypropylene bags use of various colors, *Trichoderma viride* showed maximum shelf life in milky white bags of 100-micron thickness. The *Trichoderma* fungus in the storage temperatures is less than 4°C.

## 6. Mode of action

Bio-control agents are playing an important role in controlling of plant pathogens, especially soil borne fungal pathogens. Biological control agents reduce the disease of the target crop usually by one or more of the modes of action manly antibiosis, competition, mycoparasitism, cell wall degrading enzymes and induced resistance. The indirect interaction with pathogens is competition for nutrients and space and directly with the pathogen by hyperparasitism or antibiosis [108]. Bio-control agents might directly interact with the pathogens by hyperparasitism [109], and antibiosis [110]. Bio-agents induce resistance enhanced in plants against pathogens, competitions for nutrients and spaces [111]. Various chemical compounds such as lectins during the initial contact, recognition and cell wall-degrading enzymes such as β-1,3-glucanases, chitinases, proteinases, and lipases, during the penetration [112]. In hyper-parasitism growth of bio control agent towards the target organism, coiling, final attack and dissolution of target pathogens cell wall by the activity of enzymes [86].

Mycoparasitism is one of the most important direct antagonism mechanisms that attack one fungus on another [113]. and causes complete death of fungal propagules or destruction and lysis [114]. Mycoparasitism is a complex process which involves chemotrophic growth, recognition and coiling, the interaction of hyphae and secretion of specific lytic enzymes [113]. *Trichoderma* hyphae, initial recognition and wind around the pathogen's hyphae by forming a hook, the appressorium permeates into the pathogen cell, and chitin is broken down by enzymes such as chitinase and glucanase [109]. The fungal cell walls contain chitin and glucan are the major constituents of many fungal cells [115]. *Trichoderma* strains have antagonistic potential and are mainly characterized by their ability to secrete enzymes such as chitinases, glucanases, and proteases that hydrolyze the cell walls of pathogens [116]. Chemotrophic response fungus induces the released the cell wall degrading enzymes from *Trichoderma viz.* β-1, 3 glucanase, proteases, lipases and chitinases [117]. The role of proteases in biocontrol of *Botrytis cinerea* by *T. harzianum* [57]. In mycoparasitism has been attributed to the role of chitinases [118]. The proteases reduced the activities of the pathogen enzymes exo- and endo polygalacturonase, pectin methylesterase, pectate lyase, chitinase, cutinase, and β-1,3-glucanase that are essential during host infection. *Trichoderma* hyphae contact and start coiling around the attachment of hyphae [119]. The hyphae grow along the host hyphae and secrete different lytic enzymes such as glucanase, chitinase and pectinase that are involved in mycoparasitism and ultimately degeneration of the target fungus [119]. *Trichoderma* produces low molecular weight compounds that have antifungal and antibacterial properties, these substances inhibit cell wall synthesis [118]. *Trichoderma* hyphae release antibiotic compounds which penetrate the pathogen's hyphae and inhibition of host cell wall synthesis [59, 120]. The mycoparasitism of *Trichoderma* spp. towards *Pythium ultimum* and *Sclerotium rolfsii* [18]. The parasitism of *Rhizoctonia solani* hyphae by the *Trichoderma virens* in controlling citrus seedling disease [121, 122]. *Trichoderma* species such as *T. atroviride*, *T. virens,* and *T. reesei* have ability of mycoparasitism [7, 123]. *Trichoderma harzianum* is excellent mycoparasitic activity against *Rhizoctonia solsni* [11] and also involve chitinase and β-1, 3 glucanase [1]. *Trichoderma* spp. is known to produce antimicrobial metabolites that act via hyperparasitism [10].

Antibiosisis is the condition of antagonistic to the suppression of pathogenic microorganisms due to toxic compounds (antibiotics). Antibiotic is a secondary metabolite with a low molecular weight that is harmful to the other microorganisms at a low concentration [124]. The antibiotic is produced by bio-control agents and is the main contributing mechanism under soil conditions [125]. Soil-borne microorganisms have different strains of *Trichoderma* species [5]. Secondary metabolites secreted in situ and effects against pathogens at low amounts [110].

Competition is the form of microbial in soils and living plant surfaces for nutrient limited environments [126]. Bio- control agents and pathogens compete with one another for the nutrients and space in the environment. The competition is considered to be an indirect interaction between the pathogen and bio control agent [127]. The competition for nutrients of bio-control agents fights for the essential micronutrients such as iron and manganese in soils. The bio control agents have more efficient for utilizing micro-nutrient uptake for the substances than the pathogens [128]. Iron competition is a limiting factor in alkaline soils for microbial growth and development [129]. Siderophore is low-molecular weight chelators with is a very high and specific affinity for Fe called siderophores [130]. *Trichoderma* spp. produces highly efficient low molecular weight ferric iron chelators termed siderophores that stop the growth of other fungi [131]. Siderophore is a chealate of the Fe ions that bind and take up the Siderophore-Fe-complex and making iron unavailable to the pathogen [132]. *Trichoderma* species as bio-control antagonists release siderophores that chelate iron ($Fe3+$) prevent the growth and development of fungal pathogens [90]. *Trichoderma asperellum* producing iron-binding siderophores controls *Fusarium* wilt [133].

Induce resistance is indirect mechanism in host physio-biochemical pathways that trigger defense cascades inside the plants and lead to suppression of disease development. Induced defense mechanisms involve the production of reactive oxygen species, phytoalexins, phenolic compounds, pathogenesis-related proteins, physical barriers [134]. The role of *Trichoderma* in plants defense as involved in induced immunity. Concept of induces resistance on cucumber seedling disease with *T. harzianum* [43]. The roots are recognized fungal-derived molecules that changes occur locally and systemically in gene expression, increasing salicylic acid, jasmonic acid, and phytoalexin levels in plants. The induced resistance is enhanced against infections by a pathogen in the plant without direct antagonistic interaction with the pathogen [135, 136]. *Trichoderma* application cause induces resistance against the diseases in plants and provides long-term protection [137]. *Trichoderma* in the rhizosphere can protect plants against aerial pathogen infections, through the induction of resistance via a hypersensitive response (HR), systemic acquired resistance (SAR) and induced systemic resistance (ISR) in plants [10]. Induced resistance was demonstrated through the induction of *Trichoderma* against foliage disease of beans caused by *Colletotrichum lindemuthianum* and *Botrytis cinerea* [138]. *Trichoderma* produces several metabolites that act as elicitors of plant results in the synthesis of phytoalexins, PR proteins that increase in resistance against several plant pathogens [139] and in abiotic conditions [10].

## 7. Diseases management

*Trichoderma* spp. and *Gliocladium* spp. were the first bio-control agents that effectively manage plant pathogens such as *Sclerotium rolfsii*, *Rhizoctonia solani*, and *Fusarium solani,* cause diseases on groundnut, bean, and apple, respectively

[94, 138, 140]. *Trichoderma viride, Trichoderma harzianum* and *Trichoderma virens* are being successfully used for the control of diseases such as foot rots, root rots, damping off, collar rots and *Fusarium* wilts of horticultural crops. *Trchoderma* are effective against foliar and soil borne plant pathogens [141]. The talc based formulations of *Trichoderma* manage several soil-borne diseases of various crops by seed treatment at 4 g–5 g/kg seed. Soil borne plant pathogens are successfully manage through seed coating, furrow application and root dip of seedlings. Successfully managed *S. rolfsii* and *Pythium* spp. on radish and pea by seed coating of *T. hamatum* as reported [142]. *Trichoderma harzianum* application in the field with wheat bran colonized rapidly in the soil and inhibits the *Rhizoctonia solani* and *S. rolfsii* in beans [26, 143]. *Trichoderma* spp. has potential in controlling wilt and damping-off diseases caused by *Fusarium* sp. and *Rhizoctonia solani* [28, 121]. *Pythium,* and *Phytophthora, Rhizoctonia, Fusarium and Sclerotium*, spp. are soil borne plant pathogens causing diseases in several crops, the diseases manage through *Trichoderma* spp. as reported [22]. *Trchoderma* has the potential to manage fungal and nematode diseases as well as host defense inducing ability in plants [64]. Nano-particles (Ag and Au) of *Trichoderma asperellum* showed antifungal activity at different concentrations, the maximum radial growth inhibition was observed at 200 ppm against *Rhizoctonia solani* as compared with the chemical [23].

## 8. Conclusion

*Trichoderma* is an excellent bio-control agent and also plant growth promotion. *Trichoderma spp.* serves as a bio-control agent due to effective against a large number of soil-borne plant pathogenic fungi and effects on some root nematodes. *Trichoderma* isolates improved the growth of root length, plant height, roots and shoots fresh weight and dry weight of plants. *Trichoderma* is effective against abiotic stresses and provides tolerance to the abiotic stresses and increases fertilizer used efficiencies. *Trichoderma* species have the potential to produce several enzymes that can degrade the cell wall and release several fungi toxic substances that can inhibit the growth of the fungal pathogens. A single application of *Trichoderma* give the long term efficacy as induces resistance against plants diseases. Applications of *Trichoderma* in nurseries and fields to suppress the soil-borne inoculums. The bio-efficacy is the quality of *Trchoderma* products to ensure the betterment. *Trichoderma* formulations can be applied to dry seed treatment or seed biopriming for the control of several soil-borne diseases. *Trichoderma* has high efficiency and aggressiveness against *Macreophomina phaseolina, Aspergillus niger* and *Meloidogyne incognita. Trichoderma harzianum* has the best antagonism to *Macrophomina phaseolina* and *Aspergillu. niger*, while the *Trichoderma asperellum* is efficient in the reduction of nematode. *Trichoderma* has no harmful effects on the environment and non-target organisms and can be applied to most food crops. Bio-control technologies have disease control of the crops and these technologies minimize the usage of harmful chemical pesticides. *Trichoderma* strains play an important role in the bioremediation of soil that is contaminated with pesticides and herbicides having the ability to degrade.

## Conflict of interest

"The authors declare no conflict of interest."

## Author details

Amar Bahadur[1] and Pranab Dutta[2*]

1 College of Agriculture, Tripura, Agartala, Tripura, India

2 School of Crop Protection, College of Post Graduate Studies in Agricultural Sciences, Central Agricultural University, Umiam, Meghalaya, India

*Address all correspondence to: pranabdutta74@gmail.com

## References


[1] Harman GE. Myths and dogmas of Bio control: Changes in the perceptions derived from research on *Trichoderma harzianum* T-22. Plant Diseases. 2000;**84**:377-393

[2] Puyam A. Advent of *Trichoderma* as a bio-control agent. A Review Journal of Applied and Natural Science. 2016;**8**(2):1100-1109

[3] Yedidia I, Shoresh M, Kerem Z, Benhamou N, Kapulnik Y, Chet I. Concomitant induction of systemic resistance to *Pseudomonas syringae* pv. *lachrymans* in cucumber by *Trichoderma asperellum* (T203) and accumulation of phytoalexins. Applied and Environmental Microbiology. 2003;**69**:7343-7353

[4] Chet I. Biotechnology in Plant Disease Control. New York: Wiley-Liss; 1993. 373 p

[5] Benitez MT, Ana M, Rincón M, Carmen LA, Codón C. Biocontrol mechanisms of *Trichoderma* strains. International Journal of Microbiology. 2004;7:249-260

[6] Alexander M. Introduction to Soil Microbiology. New York and London: Wiley J & Sons; 1961

[7] Sahebani N, Hadavi N. Biological control of the rootknot nematode *Meloidogyne javanica* by *Trichoderma harzianum*. Soil Biology and Biochemistry. 2008;**40**:2016-2020

[8] Singh UB, Malviya D, Wasiullah SS, Pradhan JK, Singh BP, Roy M, et al. Bio-protective microbial agents from rhizosphere eco-systems triggering plant defense responses provide protection against sheath blight disease in rice (*Oryza sativa* L.). Microbiological Research. 2016a;**192**:300-312

[9] Singh UB, Deepti M, Shailendra WS, Mohd I, Neelam P, Manzar A, et al. Compatible salt-tolerant rhizosphere microbemediated induction of phenylpropanoid cascade and induced systemic responses against Bipolaris sorokiniana (Sacc.) Shoemaker causing spot blotch disease in wheat (Triticum aestivum L.). Applied Soil Ecology. 2016b;**108**:300-306

[10] Harman GE, Howell CR, Viterbo A, Chet I, Lorito M. *Trichoderma* species-opportunistic, avirulent plant symbionts. Nature Reviews Microbiology. 2004;**2**:43-56

[11] Altomare C, Norvell WA, Bjorkman T, Harman GE. Solubilization of phosphate and micro nutrients by the plant growth promoting fungus *Trichoderma harzianum* Rifai. Applied and Environmental Microbiology. 1999;**65**:2926-2933

[12] Coley-Smith JR, Ridoμt CJ, Mitchell CM, Lynch JM. Control of button rot disease of lettuce (*Rhizoctonia solani*) using preparations of *Trichoderma viride T. harzianum*. Plant Pathology. 1991;**40**:359-366

[13] Sivan A, Elad Y, Chet I. Biological control effects of a new isolate of *Trichoderma harzianum* on *Pythium aphanidermatum*. Phytopathololgy. 1984;**74**:498-501

[14] Marra R, Ambrosino P, Carbone V, Vinale F, Woo SL, Ruocco M. Study of the three-way interaction between *Trichoderma atroviride,* plant and fungal pathogens by using a proteomic

approach. Current Genetics. 2006;**50**:307-321

[15] Labudova I, Gogorova L. Biological control of phytopathogenic fungi through lytic action of *Trichoderma* species. FEMS Microbiology Letters. 1988;**52**:193-198

[16] Dolatabadi HK, Goltapeh EM, Mohammadi N, Rabiey M, Rohani N, Varma A. Biocontrol potential of root endophytic fungi and trichoderma species against fusarium wilt of lentil under in vitro and greenhouse conditions. Agricultural Science and Technology. 2012;**14**:407-420

[17] Harman GE. Overview of mechanisms and uses of *Trichoderma* sp. Phytopathology. 2006;**96**:190-194

[18] Papavizas GC. *Trichoderma* and *Gliocladium:* biology, ecology, and potential for biocontrol. Annual Reviews of Phytopathology. 1985;**23**:23-54

[19] Rajkonda JN, Sawant VS, Ambuse MG, Bhale UN. Inimical potential of Trichoderma species against pathogenic fungi. Plant Sciences Feed. 2011;**1**(1):10-13

[20] Chet I, Hadar Y, Efad Y, Katan J, Henis Y. Biological control of soil-borne pla nt pathogens by Trichoderma harzianum. In: Schippers D, Games W, editors. Soil-Boren Plant Pathogens. London: Academic Press; 1979. pp. 585-592

[21] Conway KE. Use of fluid drilling gels to deliver biological control agents to soil. Plant Diseases. 1986;**70**:835-839

[22] Cook RJ, Baker KF. The Nature and Practice of Biological Control of Plant Pathogens. St. Paul, MN, USA: APS Books; 1983. p. 539

[23] Das A, Dutta P. Antifungal activity of biogenically synthesized silver and gold nanoparticles against sheath blight of rice. Journal of Nanoscience and Nanotechnology. 2021;**21**(6):3547-3555

[24] Das BC, Das BK, Pranab D, Sarmah DK. Bioformulation of *Trichoderma harzianum* Rifai for management of soyabean stem-rot caused by *Rhizoctonia solani* Kuhn. Journal of Biological Control. 2006;**20**(1):57-64

[25] Das BC, Hazarika DK. Biological management of sheath blight of rice. Indian Phytopathology. 2000;**53**: 433-435

[26] Elad Y, Chet IK. *Trichoderma harzianum:* A biocontrol agent effective against Sclerotium rolfsii and *Rhizoctonia solani*. Phytopathology. 1980;**70**:119-121

[27] Henis Y, Adams PB, Lewis JA, Papavizas GC. Penetration of sclerotia and *Sclerotium rolfsii* by T*richoderma* spp. Phytopathology. 1983;**73**:1043-1046

[28] Kamala T, Devi Indira S. Biocontrol properties of indigenous *Trichoderma* isolates from North-east India against *Fusarium oxysporum* and *Rhizoctonia solani*. African. Biotechnology. 2012;**11**(34):8491-8499

[29] Mathur SB, Sarbhoy AK. Biological control of Sclerotium root rot of sugar beet. Indian Phytopathology. 1978;**31**:36-367

[30] Ahmad JS, Baker R. Rhizosphere competence of *Trichoderma harzianum*. Phytopathology. 1987;**77**:182-189

[31] Lewis JA, Lumsden RD. Biocontrol of damping-off of greenhouse-grown crops caused by *Rhizoctonia solani* with a formulation of *Trichoderma* spp. Crop Protection. 2001;**20**:49-56

[32] Prasad RD, Rangeshwaran R, Anuroop CP, Phanikumar PR. Bioefficacy and shelf life of conidial and chlamydospore formulation of *Trichoderma harzianum* in talc formulation. Indian Journal of Agricultural Sciences. 2002a;**80**:930-932

[33] Prasad RD, Rangeshwaran R, Hegde SV, Anuroop CP. Effect of soil and seed application of *Trichoderma harzianum* on pigeonpea wilt caused by *Fusarium udum* under field conditions. Crop Protection. 2002b;**21**:293-297

[34] Samuels GJ. *Trichoderma*: A review of biology and systematic of the genus. Mycological Research. 1996;**100**:923-935

[35] Sivan A, Chet I. Biological control of *Fusarium* spp. in cotton, wheat and muskmelon by *Trichoderma harzianum*. Journal of Phytopathology. 1986;**16**:39-47

[36] Sivan A, Chet I. Biological control of Fusarium crown rot of tomato by Trichoderma harzianum under field conditions. Plant Dis. 1987;**71**:589-592

[37] Batta YA. Postharvest biological control of apple gray mold by *Trichoderma harzianum* Rifai formulated in an invert emulsion. Crop Protection. 2005;**23**(1):19-26

[38] Pranab D, Das BC. Management of collar rot of tomato by *Trichoderma* spp and chemicals. Indian Phytopathology. 2002;**55**(2):235-237

[39] Sharma P, Sharma M, Raja M, Shanmugam V. Status of *Trichoderma* research in India: A review. Indian Phytopathology. 2014;**67**:1-19

[40] Nelson ME, Powelson ML. Biological control of grey mold of snap beans by *Trichoderma hamatum*. Plant Diseases. 1988a;**72**:727-729

[41] Nelson ME, Powelson ML. Biological control of gray mold of snap beans by *Trichoderma hamatum*. Plant Disease. 1988b;**72**:727-729

[42] Elad Y, Zimand G, Zaqs Y, Zuriel S, Chet I. Use of *Trichoderma harzianum* in combination or alternation with fungicides to control cucumber grey mould (Botrytis cinerea) under commercial greenhouse conditions. Plant Pathology. 1993;**42**:324-332

[43] Yedidia I, Benhamou N, Chet I. Induction of defense responses in cucumber plants (Cucumis sativus L) by the biocontrol agent *Trichoderma harzianum*. Applied and Environmental Microbiology. 1999;**65**:1061-1070

[44] Elad Y, Kirshner B, Nitzani Y, Sztejnberg A. Management of powdery mildew and gray mold of cucumber by Trichoderma harzianum T39 and Ampelomyces quisqualis AQ10. Biological Control. 1998;**43**:241-251

[45] Elad Y. Biological control of foliar pathogens by means of *Trichoderma harzianum* and potential modes of action. Crop Protection. 2000;**19**:709-714

[46] Elad Y. Mycoparasitism. In: Singh KK, RP USS, editors. Pathogenesis and Host Specificity in Plant Diseases: Histopathological, Biochemical, Genetic and Molecular Basis. Vol. vol. 2: Eukaryotes Pergamon. Oxford: UK: Elsevier Science Ltd; 1995. pp. 289-307

[47] O'Neill TM, Elad Y, Shtienberg D, Cohen A. Control of grapevine grey mould with *Trichoderma harzianum* T39. Biocontrol Science and Technology. 1996;**6**:139-146

[48] Krishnamurthy YL, Shashikala J. Inhibition of aflatoxin B1 production of *Aspergillus flavus*, isolated from soybean

seeds by certain natural plant products. Letters in Applied Microbiology. 2006;**43**:469-474

[49] Kloepper JW, Tuzun S, Kuc J. Proposed definitions related to induced resistance. Biocontrol Science and Technology. 1992;**2**:349-351

[50] Sequeira L. Mechanisms of induced resistance in plants. Annual Review of Microbiology. 1983;**37**:51-79

[51] Paulitz TC, Matta A. The role of the host in biological control of diseases In: Albajes R, Gullino ML, van Lenteren JC, Elad Y (eds), Kluwer Academic Publisher, Wageningen, the Netherlands, 1999. p. 394-410

[52] Lifshitz R, Dupler M, Elad Y, Baker R. Hyphal interactions between a mycoparasite, Pythium nunn, and several soil fungi. Canadian Journal of Microbiology. 1984;**30**:1482-1487

[53] Howell C, Stipanovic R, Lumsden R. Antibiotic production by strains of *Gliocladium virens* and its relation to bio-control of cotton seedling diseases. Biocontrol Science and Technology. 1993;**3**:435-441

[54] De Meyer G, Bigirimana J, Elad Y, Höfte M. Induced systemic resistance in *Trichoderma harzianum* T39 biocontrol of *Botrytis cinerea*. European Journal of Plant Pathology. 1998;**104**:279-286

[55] Kapat A, Zimand G, Elad Y. Biosynthesis of pathogenicity hydrolytic enzymes by Botrytis cinerea during infection of bean leaves and in vitro. Mycological Research. 1998b;**102**:1017-1024

[56] Kapat A, Zimand Y, Elad Y. Effect of two isolates of *Trichoderma harzianum* on the activity of hydrolytic enzymes produced by Botrytis cinerea. Physiological and Molecular Plant Pathology. 1998a;**52**:127-137

[57] Elad Y, Kapat A. Role of *Trichoderma harzianum* protease in the biocontrol of *Botrytis cinerea*. European Journal of Plant Pathology. 1999;**105**:177-189

[58] Haab D, Hagspiel K, Szakmary K, Kubicek CP. Formation of the extracellular protease from *Trichoderma reesei* QM 9414 involved in cellulase degradation. Journal of Biochemical Technology. 1990;**16**:187-198

[59] Rodriguez-Kabana R, Kelley WD, Curl EA. Proteolytic activity of *Trichoderma viridein* mixed culture with *Sclerotium rolfsii* in soil. Canadian Journal of Microbiology. 1978;**24**:487-490

[60] Ilias GNM, Rahman MA, Molla AH, et al. Composting of municipal garbage by using *Trichoderma*- a new approach in context of Bangladesh. Bangladesh Journal of Genetic Biotechnology. 2005;**6**(1&2):75-78

[61] Yang ZS, Li GH, Zhao PJ, Zheng X, Luo SL, Li L, et al. Nematicidal activity of Trichoderma spp. and isolation of an active compound. World Journal of Microbiology and Biotechnology. 2010;**26**:2297-2302

[62] Szabó M, Csepregi K, Gálber M, Virányi F, Fekete C. Control plant-parasitic nematodes with *Trichoderma* species and nematode-trapping fungi: The role of chi18-5 and chi18-12 genes in nematode egg-parasitism. Biological Control. 2012;**63**:121-128

[63] Arora DK, Elander RP, Mukerji KG. Handbook of Applied Mycology. New York: Marcel Dekker; 1992

[64] Harman GE, Kubicek CP, editors. *Trichoderma* and *Gliocladium*, Vol. 1 & 2. London: Taylor and Francis; 1998 278 p

[65] Harman GE, Mattick LR, Nash G, Nedrow BL. Stimulation of fungal spore germination and inhibition of sporulation in funga l vegetation thalli by fatty acids and the in volatile peroxidation products. Canadian Journal of Botany. 2011;**58**:1541-1547

[66] Eyal J, Baker CP, Reeder JD, Devane WE, Lumsden RD. Large-scale production of chlamydospores of *Gliocladium virens* strain GL-21 in submerged culture. Journal of Industrial Microbiology & Biotechnology. 1997;**19**:163-168

[67] Harman GE, Jin X, Stasz TE, Peruzzotti G, Leopold AC, Taylor AG. Production of conidial biomass of *Trichoderma harzianum* for biological control. Biological Control. 1991;**1**:23-28

[68] Jin X, Harman GE, Taylor AG. Conidial biomass and desiccation tolerance of *Trichoderma harzianum* produced at different medium water potentials. Biological Control. 1991;**1**:237-243

[69] Kirk PM, Cannon PF, David JC, Stalpers JA. Dictionary of the fungi. 10th ed. Wallingford: CAB International; 2008

[70] Pranab D, Das BC. Biomass production of *Trichoderma harzianum* using various common wastes. New Agriculturist. 1997;**8**(1):85-88

[71] Pranab D, Islam M. Evaluation of organic substrates for growth and multiplication of indigenous biocontrol agent. Journal Nature and Environment. 2010;**2**(3):12-16

[72] Selvakumar R, Srivastava KD. Management of Spot Blotch of Wheat Using Bio Control Agent. New Delhi: National Conference on Swadeshi Vighan, held at; December, 2000. pp. 26-28

[73] Upadhyay JP, Mukhopadhyay AN. Biological bioefficacy of *Trichoderma harzianum* control of *Sclerotium rolfsii* by *Trichoderma harzianum* in Sugarbeet. Tropical Pest Disease Management. 1986;**32**:215-220

[74] Jogani V, John P. Evaluation of different application methods of *Trichoderma harzianum* (Rifai) against Fusarium wilt of tomato. Crop Research. 2014;**48**(1-3):76-79

[75] Jeyarajan R. Prospects of indigenous mass production and formulation of *Trichoderma*. In: Rabindra RJ, Ramanujam B, editors. Current Status of Biological Control of Plant diseases using antagonistic organisms in India. Bangalore: Project Directorate of Biological Control; 2006. pp. 74-80

[76] Zaid WN, Singh US. Development of improved technology for the mass multiplication and delivery of Fungal (*Trichoderma*) and Bacterial (*Pseudomonas*) bio agents. Journal of Mycology and Plant Pathology. 2004;**34**:732-745

[77] Sankar P, Jeyarajan R. Biological control of *Trichoderma harzianum*-a potential biocontrol agent for tobacco damping off. Indian Research. 1996;**12**:26-35

[78] Prasad RD, Rangeshwaran R. A modified liquid medium for mass production of *Trichoderma* by fermentation process, pp. 26. In: Abstracts of National symposium on Eco-friendly approaches in the management of plant diseases. Shimoga, Karnataka, India; December 1998. pp. 22-24

[79] Papavizas GC, Lewis JA. Effect of *Gliocladium* and *Trichoderma* on damping off and blight of snap bean caused by *Sclerotium rolfsii* in the greenhouse. Plant Pathology. 1989;**38**:277-286

[80] Kumar S. *Trichoderma*: A biological weapon for managing plant diseases and promoting sustainability. International Journal of Agricultural Science and Veterinary Medicine. 2013;**1**(3):1-16

[81] Lumsden RD, Lewis JA. Selection, production, formulation and commercial use of plant disease biocontrol fungi for improving plant growth. In: Whipps JM, Lumsden RD, editors. Biotechnology of Fungi for Improving Plant Growth. Cambridge: Cambridge University Press; 1989. pp. 171-190

[82] Jeyarajan R, Ramakrishnan G, Dinakaran D, Sridar R. Development of products of *Trichoderma viride* and *Bacillus subtilis* for bio-control of root rot diseases. In: Dwivedi BK, editor. Biotechnology in India. Allahabad: Bioved Research Society; 1994. pp. 25-36

[83] Lewis JA, Papavizas GC. Biocontrol of plant diseases: the approach for tomorrow. Crop Protection. 1991;**10**:95-105

[84] Sawant IS, Sawant SD. A simple method for achieving high cfu of *Trichoderma harzianum* on organic wastes for field applications. Indian Phytopathology. 1996;**49**:185-187

[85] Mukhopadhyay AN, Shrestha SM, Mukherjee PK. Biological seed treatment for control of soil borne plant pathogens. FAO Plant Protection Bulletin. 1992;**40**:21-30

[86] Tewari AK. Biological Control of chickpea wilt- complex using different formulations of *Gliocladium virens* through seed treatment [thesis]. Pantnagar, India: GB Pant University of Agriculture and Technology; 1996. p. 167

[87] Mukherjee PK, Mukhopadhyay AN. In situ Mycoparasitism of *Gliocladium virens* on *Rhizoctonia solani*. Indian Phytopathology. 1995;**48**(1):101-102

[88] Singh D, Maheshwari VK. Biological seed treatment for the control of loose smut of wheat. Indian Phytopathology. 2001;**54**(4):457-460

[89] Dutta P, Das BC, Hazarika DK. Integrated management of soybean stem rot. Journal of Biological Control. 2000;**14**(1):67-69

[90] Naher L, Yusuf UK, Ismail A, Hossain K. *Trichoderma* spp.: A biocontrol agent for sustainable management of plant diseases. Pakistan Journal of Botany. 2014;**46**(4):1489-1493

[91] Kumar S, Thakur M, Rani A. *Trichoderma*: Mass production, formulation, quality control, delivery and its scope in commercialization in India for the management of plant diseases. African Journal of Agricultural Research. 2014;**9**(53):3838-3852

[92] Singh US, Zaidi NW. Current status of formulation and delivery of fungal and bacterial antagonists for disease management in India. In: Rabindra RJ, Hussaini SS, Ramanujam B, editors. Microbial Biopesticide Formulations and Application. Bangalore: Project Directorate of Biological Control; 2002. pp. 168-179

[93] Vasudevan P, Kavitha S, Priyadarisini VB, Babujee L, Gnanamanickamm SS. Biological control of rice diseases. In: Gnanamanickam SS, editor. Biological Control of Crop Diseases. Newyork: Marcel Decker; 2002. pp. 11-32

[94] Baby UI, Manibhushanrao K. Fungal antagonists and VA mycorrhizal fungi for biocontrol of *Rhizoctonia solani*, the rice sheath blight pathogen. In: Manibhushanrao K, Mahadevan A, editors. Recent Developments in Bio-control of Plant Pathogens. New Delhi: Today and Tomorrow's Printers and Publishers; 1996. pp. 1-9

[95] Reddy KK, Narayana P. Efficacy of *Trichoderma viride* Against *Colletotrichum falcatum* in Sugarcane. Indian Journal of Plant Protection. 2009;**37**(1, 2):111-115

[96] Srivastava RK, Singh RK, Kumar N, Singh S. Management of *Macrophomia* disease complex in jute (*Corchorus olitorius*) by *Trichoderma viride*. Journal of Biological Control. 2010;**24**(1):77-79

[97] Samuels GJ, Dodd SL, Gams W, Castlebury LA, Petrini O. *Trichoderma* species associated with the green mold epidemic of commercially grown *Agaricus bisporus*. Mycologia. 2002;**94**:146-170

[98] Dutta P, Das BC. Control of *Rhizoctonia solani* in soybean (*Glysine max*) by Farmyard manure culture of *Trichoderma harzianum*. Indian Journal of Agricultural Sciences. 1999;**69**(8):596-598

[99] Kumar S, Upadhyay JP, Kumar S. Biocontrol of *Alternaria* leaf spot of *Vicia faba* using antogonistic fungi. Journal of Biological Control. 2005;**20**(2):247-251

[100] Khan AA, Sinha AP. Influence of different factors on the effectivity of fungal bioagents to manage rice sheath blight in nursery. Indian Phytopathology. 2005;**3**:289-293

[101] Khan AA, Sinha AP. Screening of *Trichoderma* spp. Against *Rhizoctonia solani* the causal agent of rice sheath blight. Indian Phytopathology. 2007;**60**(4):450-456

[102] Andrews JH. Biological control in the phyllosphere. Annual Review of Phytopathology. 1992;**30**:603-635

[103] Upamanya GK, Bhattacharyya A, Pranab D. Consortia of entomopathogenic fungi and biocontrol agents improve the agro ecological conditions of brinjal cultivation in Assam. 3 Biotech. 2020;**10**:450. DOI: 10.1007/s13205-020-02439-3

[104] Adekunle AT, Cardwell KF, Florini DA, Ikotun T. Seed treatment with Trichoderma species for control of damping-off of cowpea caused by *Macrophomina phaseolina*. Biocontrol Science and Technology. 2001;**11**:449-457

[105] Chaube HS, Mishra DS, Varshney S, Singh US. Bio control of plant pathogens by fungal antagonists: a historical background, present status and future prospects. Annual Review of Plant Pathology. 2003;**2**:1-42

[106] Sriram S, Palanna KB, Ramanujam B. Effect of chitin on the shelf life of *Trichoderma harzianum* in talc formulation. Indian Journal of Agriculture Sciences. 2010;**80**:930-932

[107] Sriram S, Roopa KP, Savitha MJ. Extended shelf- life of liquid fermentation derived talc formulations of *Trichoderma harzianum* with the addition of glycerol. Crop Protection. 2011;**30**:1334-1339

[108] Spadaro D, Droby S. Development of biocontrol products for postharvest diseases of fruit: the importance of elucidating the mechanisms of action of yeast antagonists. Trends in Food Science and Technology. 2016;**47**:39-49

[109] Ghorbanpour M, Omidvari M, Abbaszadeh-Dahaji P, Omidvar R, Kariman K. Mechanisms underlying the protective effects of beneficial fungi

against plant diseases. Biological Control. 2018;**117**:147-157

[110] Raaijmakers JM, Mazzola M. Diversity and natural functions of antibiotics produced by beneficial and plant pathogenic bacteria. Annual Review of Phytopathology. 2012;**50**:403-424

[111] Kohl J, Kolnaar R, Ravensberg WJ. Mode of action of microbial biological control agents against plant diseases: relevance beyond efficacy. Frontiers in Plant Science. 2019;**10**:845

[112] Vos CM, De Cremer K, Cammue BPA, De Coninck B. The toolbox of *Trichoderma* spp. in the biocontrol of *Botrytis cinerea* disease. Molecular Plant Pathology. 2015;**16**:400-412

[113] Dix NJ, Webster J. Fungal Ecology. 1st ed. London: Chapman and Hall; 1995

[114] Maloy OC. Plant Disease Control: Principles and Practice. New York: John Wiley Sons Inc.; 1993. p. 346

[115] Lam ST, Gaffney TD. Biological activities of bacteria used in plant pathogen control. In: Chet I, editor. Biotechnology in Plant Disease Control. New York: John Wiley; 1993. pp. 291-320

[116] Lopez-Mondéjar R, Ros M, Pascual JA. Mycoparasitism related genes expression of *Trichoderma harzianum* isolates to evaluate their efficacy as biological control agent. Biological Control. 2011;**56**:59-66

[117] Viterbo A, Inbar J, Hadas Y, Chet I. Plant Disease Biocontrol and Induced Resistance via Fungal Mycoparasites: The Mycota-A Comprehensive Treatise on Fungi as Experimental Systems for Basic and Applied Research. In: Environmental and Microbial Relationships. 2nd ed. Springer; 2007. p. 350

[118] Lorito M, Woo SL, D'Ambrosio M, Harman G.E, Hayes CK, Kubicek CP, Scala F. Synergistic interaction between cell wall degrading enzymes and membrane affecting compounds. Molecular Plant-Microbe Interactions. 1996;**9**:206-213

[119] Howell CR. Mechanisms employed by trichoderma species in the biological control of plant diseases: The history and evolution of current concepts. Plant Disease. 2003;**87**:4-10

[120] Toghueoa RMK, Ekea P, Gonzálezb IZ, de Aldanab BRV, Nanaa LW, Boyoma FF. Biocontrol and growth enhancement potential of two endophytic *Trichoderma* spp. from *Terminalia catappa* against the causative agent of common bean root rot (*Fusarium solani*). Biological Control. 2016;**96**:8-20

[121] Rojo FG, Reynoso MM, Sofia M, Chulze IN, Torres AM. Biological control by Trichoderma species of *Fusarium solani* causing peanut brown root rot under field conditions. Crop Protection. 2007;**26**:549-555

[122] Weindling R. *Trichoderma lignorum* as a parasite of other fungi. Phytopathology. 1932;**22**:837-845

[123] Kubicek CP, Herrera-Estrella A, Seidl-Seiboth V, Martinez DA, Druzhinina IS, Thon M. Comparative genome sequence analysis underscores mycoparasitism as the ancestral life style of *Trichoderma*. Genome Biology. 2011;**12**(4):40

[124] Fravel DR. Role of antibiosis in the biocontrol of plant diseases. Annual Review of Phytopathology. 1988;**26**:75-91

[125] Haas D, Défago G. Biological control of soil-borne pathogens by *fluorescent Pseudomonads*. Nature Reviews Microbiology. 2005;**3**:307-319

[126] Pal KK, McSpadden Gardener B. Biological control of plant pathogens. The Plant Health Instructor. 2006:1-25. DOI: 10.1094/PHI-A-2006-1117-02

[127] Lorito M, Hayes CK, Zonia A, Scala F, Del SG, Woo SL, et al. Potential of genes and gene productsfrom *Trichoderma* sp. and *Gliocladium* sp. for the development of biological pesticides. Molecular Biotechnology. 1994;**2**:209-217

[128] Nelson EB. Exudate molecules initiating fungal responses to seed seeds and roots. Plant and Soil. 1990;**129**:61-73

[129] Leong SA, Expert D. Siderophores in plant pathogen interactions. In: Kosuge T, Nester EW, editors. Plant-Microbe Interactions Molecular and Genetic Perspectives. Vol. 3. New York: McGraw-Hill; 1989. pp. 62-83

[130] Barbeau K, Zhang G, Live DH, Butler A. Petrobactin, a photoreactive siderophore produced by the oil-degrading marine bacterium Marinobacter hydrocarbonoclasticus. Journal of the American Chemical Society. 2002;**124**(3):378-379

[131] Chet I, Inbar J. Biological control of fungal pathogens. Applied Biochemistry and Biotechnology. 1994;**48**:37-43

[132] Mukhopadhyay ASN, Mukherjee PK. Biological control of plant diseases: Status in India in Biological suppression of Plant Diseases: Phytopathogens, Nematodes and weeds. In: Singh SP, Husain SS, editors. Vol. 7. 1998. pp. 1-20

[133] Segarra G, Casanova E, Avilés M, Trillas I. *Trichoderma asperellum* strain T34 controls *Fusarium* wilt disease in tomato plants in soilless culture through competition for iron. Microbial Ecology. 2010;**59**:141-149

[134] Wiesel L, Newton AC, Elliott I, Booty D, Gilroy EM, Birch PRJ, et al. Molecular effects of resistance elicitors from biological origin and their potential for crop protection. Frontiers in Plant Science. 2014;**5**:655

[135] Conrath U, Beckers GJM, Langenbach CJG, Jaskiewicz MR. Priming for enhanced defense. Annual Review of Phytopathology. 2015;**53**:97-119

[136] Pieterse CMJ, Zamioudis C, Berendsen RL, Weller DM, Van Wees SCM, AHM BP. Induced systemic resistance by beneficial microbes. Annual Review of Phytopathology. 2014;**52**:347-375

[137] Yedidia I, Benhamou N, Kapulnik Y, Chet I. Induction and accumulation of PR proteins activity during early stages of root colonization by the mycoparasite *Trichoderma harzianum* strain T-203. Plant Physiology and Biochemistry. 2000;**38**:863-873

[138] Bigirimana J, De Meyer G, Poppe J, Elad Y, Hofte M. Induction of systemic resistance on bean (*Phaseolus vulgaris*) by *Trichoderma harzianum*. Mededelingen - Faculteit Landbouwkundige en Toegepaste Biologische Wetenschappen Universiteit Gent. 1997;**62**:1001-1007

[139] Elad Y, Freeman S, Monte E. Biocontrol Agents: Mode of Action and Interaction with other Means of Control. Sevilla, Espana: IOBC; 2000. p. 24

[140] Karampour F, Okhovat M. Investigation on the effect of some isolates of Trichoderma and Gliocladium on the growth of Fusarium solani at lab conditions. In: Proceedings-of-the-11th Plant-Protection-Congress-of-Iran-28 Aug–2-Sep. Rasht: Iran Islamic Republic; 1993. p. 143

[141] Ha TN. Using Trichoderma species for biological Control of Plant Pathogens in Vietnam. Journal of the International Society for Southeast Asian Agricultural Sciences. 2010;**16**(1):17-21

[142] Harman GE, Mattick LR, Nash G, Nedrow BL. Stimulation of fungal spore germination and inhibition of sporulation in fungal vegetation thalli by fatty acids and the in volatile peroxidation products. Canadian Journal Botany. 1980;**58**:1541-1547

[143] Hadar Y, Chet I, Henis Y. Biological control of *Rhizoctonia solani* damping-off with wheat bran culture of *Trichoderma harzianum*. Journal of Phytopathology. 1979;**69**:64-68

Chapter 4

# Can Genus *Trichoderma* Manage Plant Diseases under Organic Agriculture?

*Kishor Chand Kumhar, Dalvinder Pal Singh and Anil Kumar*

## Abstract

Organic agriculture has been coming up as one of the promising segments of crop production systems in India. There are numerous reasons for it, however; human health, sustainable environment, soil health, etc. are the important ones. As per the latest information, India has about 1.5% of total cultivable land under organic agriculture. The occurrence of plant diseases in this crop production system is one of the limiting factors. For the management of plant diseases in organically grown crops, there are limited resources since there is a restriction on the use of synthetic fungicides. Under such a situation, bio-pesticides have the potency to take care of plant diseases. Although there are certain fungal and bacterial candidates well efficient in controlling diseases, genus *Trichoderma* has occupied a prestigious position among them. It is capable of managing seed and soil-borne plant diseases. Presently it is available in wettable powder (WP) and liquid formulations in variable concentrations for the application.

**Keywords:** *Trichoderma* spp., plant diseases, organic agriculture, mode of action, formulation

## 1. Introduction

The interest in organic crop production has been increasing day-by-day since the last decade because of the increased negative impact of conventional agriculture production systems. Production of crops adopting the inorganic agricultural inputs primarily agrochemicals, which has been used starting from seed treatment till harvesting of the crop for providing protective umbrella against various diseases. The frequent and injudicious application of inorganic agrochemical inputs has invited several environmental and health-related problems.

The demand for organic food commodities has been increased tremendously and hence researchers, as well as growers, have been focusing their attention towards organic crop production wherein, the use of inorganic inputs is completely avoided. However, successful crop production through this system faces various biotic and abiotic challenges too.

Organic farming was first initiated by British botanist Sir Albert Howard in 1905. However, in India, it was initiated during the late 90s. The government of India, in the year 2001, implemented the national program for organic production (NPOP) for the promotion of organic farming. Organic farming was started in several states. Sikkim has converted cent percent of the agriculture into an organic state in this country. Presently, India has about 2.3 million hectares under it [1]. with a production of about 1.70 million MT certified organic products with voluminous export of different organic food items [2].

The occurrence of plant diseases is one of the important concerns in organic agriculture. Various fungi are notable/major phytopathogenic fungi causing huge crop losses [3]. For the management of plant diseases, plant protection measures start as early as seed treatment and would continue until crop harvesting in different crops.

In the recent era, organic agriculture is getting huge popularity and adaptability owing to its several beneficial aspects. The organically produced agricultural commodities ensure pesticide-free items which are fit for human consumption. Under such a crop production system, taking care of plant diseases is one of the major challenges. There are limited plant protection options due to the restricted use of synthetic fungicides in it. It emphasises the use of on-farm inputs for almost all requirements. Under such circumstances, the application of *Trichoderma* spp. could be an ideal alternative to handle plant diseases. Since the last couple of decades, genus *Trichoderma* has occupied a reputed position among the microbes possessing disease management potency. As a biocontrol agent, the antagonistic property of *Trichoderma viride* was recognised [4] while working with the damping-off of citrus seedlings caused by *Rhizoctonia*. Various chemical fungicides were the most dominant during the green revolution era in the 60s and 70s and hence *Trichoderma* could not significantly attain popularity among the growers at the grass-root level. Due to excessive usage of chemical fungicides numerous problems related to the environment, soil and human health ecosystem, resistance development, etc. have emerged as a big challenge before everyone, including scientists, consumers, growers, and the overall associated environment. Therefore, majority has been preferring pesticide-free food commodities for the last so many years. It is good news or sign that usage of chemical fungicides has been coming down and usage of biopesticides has been going up nationally as well as internationally [5]. Currently, in India, there are several hundred manufacturers of *Trichoderma* possessing the registration certificate from the Central Insecticide Board and Registration Committee [6]. They are manufacturing and marketing the *Trichoderma*-based products with variable active ingredients, the dose of application in diversified crop avenues. Although there are four-five species of this antagonist suitable for the management of plant diseases, however only two species namely, *T. viride* and *T. harzianum* have been made available to the end-users. The plus point with this antagonist is that it can be used in a wide range of crops, such as cereals, pulses, vegetables, horticultural, and plantation crops [7–11]. Its wettable powder (WP) and liquid formulations are popular in the market; however, the first one is dominant. In addition to managing the phytopathogen, it is capable of promoting the vegetative growth of plants. Being a natural fungus, it is safer for the environment as well as consumers, capable of adjusting and performing under variable conditions. This chapter highlights various important information related to this genus.

This chapter mainly focuses on the application of *Trichoderma* spp. for the management of plant diseases. Organic crop production excludes the use of synthetic/inorganic fungicides for the control of plant diseases.

## 2. History and journey of *Trichoderma* species

Before touching its actual uses in agriculture, it becomes imperative to focus on some important aspects related to it. For the first time, Persoon introduced the name *Trichoderma* [12]. In 1865, the Tulasne brothers reported teleomorph (sexual stage) of *T. viride* Pers as *Hypocrea rufa* [13]. Till 1969, there were only one species, i.e., *T. viride* of genus *Trichoderma* [14].

In 1969, Rifai made out nine 'aggregate species' namely, (1) *T. harzianum* Rifai, (2) *T. viride,* (3) *T. hamatum* (Bonord.) Bainier, (4) *T. koningii* (Oudem.) Duché & R. Heim, (5) *T. polysporum* (Link) Rifai, (6) *T. piluliferum* J. Webster & Rifai, (7) *T. aureoviride* Rifai, (8) *T. longibrachiatum* Rifai, and (9) *T. pseudokoningii* Rifai. In the early 1990s, [15–17] identified five sections and 27 biological species within the genus *Trichoderma.* The introduction of molecular techniques contributed to the greater extent in identifying the species comparatively more precisely and hence from the late 1990s up to the year 2002, the number of *Trichoderma* species increased to 47 [18].

Till 2005, the International sub-commission on *Trichoderma/Hypocrea* listed 104 species on the basis of phylogenetic analyses [19]. Till the year 2015, there are 252 species one variety and one form. In addition, *T. neocrassum* Samuel (syn. *Hypocrea crassa* P. Chaverri & Samuels) and *T. patellotropicum* Samuels (syn. *Hypocrea patella* f. tropicaYoshim.Doi) were proposed as two new names [20].

## 3. Interaction of *Trichoderma* with phytopathogens of agricultural crops

Hyperparasitism, competition, and antibiosis are the important mechanisms of this genus through which it interacts with phytopathogenic fungi. This antagonist is well efficient to colonise in variable ecological niches [21–23].

### 3.1 Hyperparasitism

In Hyperparasitism, the *Trichoderma* directly contact with a pathogen and finally cause the death of pathogen cells death [23]. For this purpose, the *Trichoderma* synthesise cell wall degrading enzymes (CWDE) namely, cellulase, xylanase, pectinase, glucanase, lipase, amylase, arabinase, protease [24], and lytic enzymes. These enzymes degrade the pathogen cell walls, composed of chitin and glucan polysaccharides. Among the chitinolytic enzymes β-N-acetyl glucosaminidase, endochitinase, and chitobiosidase are of key importance responsible for the degradation of the cell wall of other plants pathogenic fungi, which are produced by *T. harzianum*, *T. atroviride* P. Karst, and *T. asperellum* Samuels, Lieckf. & Nirenberg [25]. Several volatile metabolites, such as 6-n-pentyl-2H- -pyran-2-one (6-PAP), are produced by *Trichoderma* for plant protection [26]. The β-1,3- and β-1,6-glucanases enzymes determine the hyperparasitic capability of *Trichoderma* to react on species of *Phytophthora* sp. and *Pythium* [27, 28]. Proteolytic enzymes, namely, endo- and exoproteases of *Trichoderma* responsible for enzyme secretion for the control of *Botrytis cinerea*, *Rhizoctonia solani,* and *Fusarium culmorum* [29, 30]. Ceratin cellulase enzymes, such as exo-β-1,4-glucanases, endo-β-1,4-glucanases, and β-glucosidases, produced by antagonists also play a significant role in hyperparasitism [31].

### 3.2 Competition

Competition of *Trichoderma* spp. with plant pathogens is another mode of interaction that helps in the control of plant diseases. This phenomenon takes place for utilisation of nutrients, occupying ecological position or infection sites on plant roots. Certain *Trichoderma* strains produce siderophores, i.e., iron-chelating compounds. Through siderophores, antagonist traps iron from the associated environment and creates nutrient deficiency due to which the growth of pathogenic fungi, such as *B. cinerea,* is hampered. The *Trichoderma* creates an acidic environment that has a negative effect on the growth and development of pathogenic fungi and aggressively colonises the host plant's roots due to the enhanced activity of the hydrophobins [9].

### 3.3 Antibiosis

It is a biological interaction between two or more organisms that is unfavourable to at least one of them; it can also be an antagonistic association between an organism and the metabolic substances produced by another. Antibiosis in relation to *Trichoderma* fungi is a specific mechanism of antagonistic interactions with other plant pathogenic fungi. This phenomenon is based on the generation of secondary metabolites, which exhibit an inhibitory or lethal effect on a parasitic fungus. From the genus *Trichoderma*, over 180 secondary metabolites have been characterised so far, representing different classes of chemical compounds [32, 33]. Such compounds can be divided into volatile antibiotics, water-soluble compounds, and peptaibols. Volatile antibiotic, (6PAP (6-pentyl-α-pyrone) produced by *T. viride*, *T. harzianum*, and *T. koningii*, plays a major role in the biocontrol of *Botrytis cinerea*, *R. solani*, and *Fusarium oxysporum.*

Another category of antibiotics is Peptaibols. They are polypeptide antibiotics comprising of 500–2200 Da, rich in non-proteinogenic amino acids, specifically alfa-aminoisobutyric acid, and their characteristic attributes include the presence of N-acetylated ends and C-end amino alcohols. Peptaibols exhibit potent activity against a number of fungi (**Table 1**).

| *Trichoderma* **spp.** | **Produced Peptibols antibiotics** | **Reference** |
|---|---|---|
| *T. viride* | trichotoxins A and B, trichodecenins, trichorovins, and trichocellins | [34] |
| *T. harzianum* | trichorzianins A and B, trichorzins HA and MA | [34] |
| *T. longibrachiatum* | tricholongins BI and BII, and longibrachins | [34] |
| *T. koningii* | trichokonins | [34] |
| *T. atroviride* | atroviridins A-C and neoatroviridins A-D | [34] |
| *T. koningii, T. harzianum, T. aureoviride, T. viride, T. virens* | koningins, viridin, dermadin, trichoviridin, lignoren, and koningic acid | [33] |
| *T. lignorum, T. virens (G. virens), T. viride, and T. hamatum* | mycotoxin | [33] |
| *T. harzianum* | harzianum A | [35] |

**Table 1.**
*Peptibols antibiotics of* Trichoderma *spp. to control phytopathogens.*

In *Trichoderma* fungi, the activity of antibiotics is enhanced with the activity of lytic enzymes. Their joint activity provides a higher level of antagonism compared to the activity of either enzymes or antibiotics alone [36]. Researchers found preliminary degradation of the cell walls in *B. cinerea* and *F. oxysporum* by lytic enzymes; it facilitated easier penetration of antibiotics to pathogen cells [8].

## 4. Role of *Trichoderma* spp. as a biological control agent

*Trichoderma* spp. are the ideal option for safer management of phytopathogens. *T. harzianum* and *T. viride* have been occupied the prime position and its highly exploited biological control agent for controlling soil-borne fungal diseases.

## 5. Current status of *Trichoderma* formulations

There is a list of about 970 registered manufacturers with the Central Insecticide Board & Registration Committee (CIB & RC) for the manufacturing and marketing of biopesticides in India, out of which 558 are involved in *Trichoderma* spp. (**Figure 1**). The different formulations of this antagonist comprised of wettable powder (WP), wettable suspension (WS), aqueous suspension (AS), and liquid. The *Trichoderma* spp. formulations have been manufactured and marketed by private companies, government organisations, and NGOs. Presently, wettable powder (WP) and liquid formulations are available for use by end-users; however, WP formulation in different concentrations (0.5 to 6.0%) is dominant (**Figure 2**). The liquid formulations contribute nearly 2% of the formulations.

## 6. Management through seed treatment with *Trichoderma* spp.

Seed treatment is an important and economical practice to manage the seed-borne phytopathogens at the initial stage with the minimum cost. For this purpose, a small quantity of desired input is required and this practice is very easy in handling. For seed treatment, the formulation of *T. harzianum* and *T. viride* can be used. Seedlings of vegetable crops can be treated with *T. viride* and *T. harzianum* just before transplanting into the main field (**Table 2**).

## 7. Management through soil treatment with *Trichoderma* spp.

Soil residential phytopathogens, such as *Pythium*, *Phytophthora*, *Fusarium*, *Rhizoctonia*, *Sclerotinia*, *Macrophomina,* play an important role in the development of various soil-borne diseases, such as root rot, wilt, and seedling rot of various crops. Management of such fungal phytopathogens can be achieved through the application of *T. harzianum* and *T. viride* through soil drenching with water, enriched farmyard manure, and vermicompost. There are several species of *Trichoderma*, however *T. harzianum*, *H. lixii*, *T. atroviride*, *H. atroviridis*, *T. asperellum,* and *T. virens* are potential biocontrol agents against phytopathogens [37]. Application of *T. viride* enriched FYM (5 kg/plant) to two-year guava saplings at the basin near the root zone resulted in decreased wilt incidence and better plant

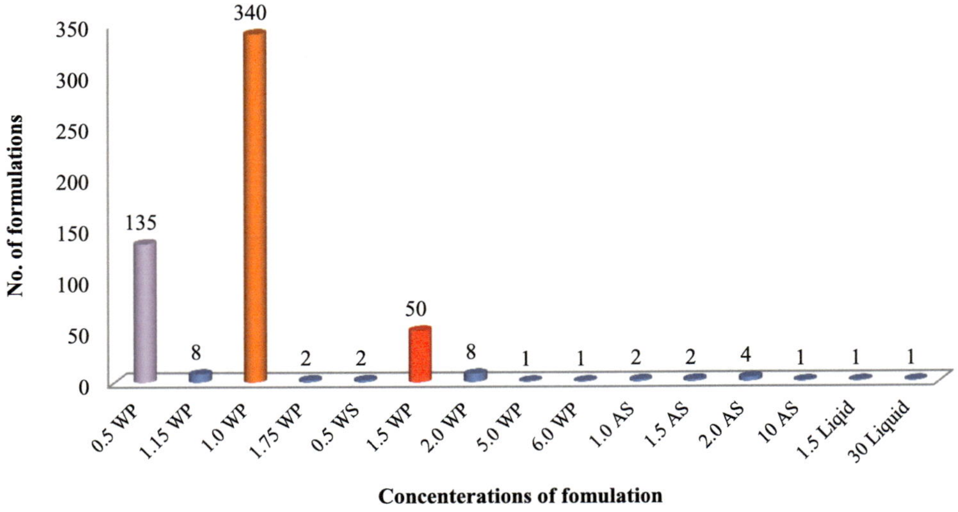


**Figure 1.**
*Number of manufacturers of different* Trichoderma *formulations.*

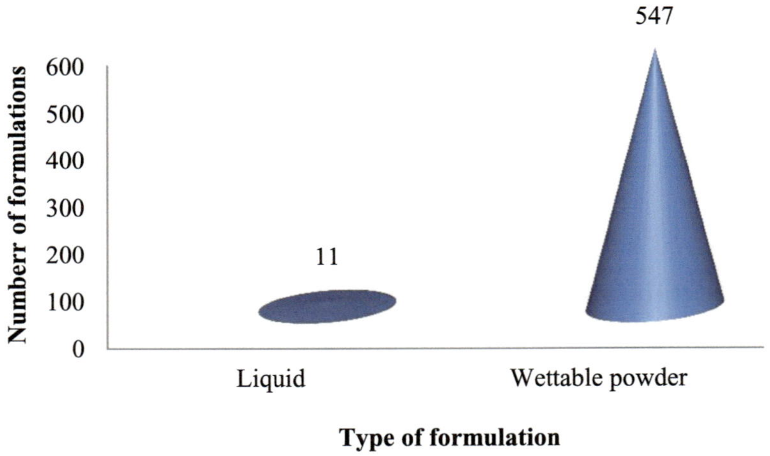


**Figure 2.**
*Status of* Trichoderma *WP and liquid formulation manufacturers.*

growth in terms of stem girth [38]. *T. harzianum* when applied (50 g/vine) in black pepper field it had effectively managed the foot rot disease. For the management of pomegranate wilt bio-formulation of *T. viride* (0.1 and 0.2%) was found significantly superior over the control. Soil application of *T. harzianum* plus *T. virens* is efficient in managing stem bleeding disease of coconut [39]. Five monthly applications of *T. harzianum* (50 g/plant) in bananas reduced 50% of the vascular discoloration index of *Fusarium* wilt disease and increased the yield [40]. *T. asperellum* found inhibitory against *Pythiumaph anidermatum*, *Pythium debaryanum*, *Sclerotium rolfsii,* and *S. rolfsii*, *Fusarium oxysporum* f.sp. *lycopersici* and *Alternaria solani*. Application of *Trichoderma* (@20 kg/ha) along with 2.0 tonnes castor cake/ha reduced nematode population and increased yield in pomegranate.

Soil application of silver nanoparticles synthesised from *T. asperellum* resulted in complete control of *Fusarium* wilt in cv. Grand Nain [41]. *T. harzianum* talc

| Input | Crop | Adopted by |
|---|---|---|
| *Trichoderma* | Chilli seed treatment | Uttarakhand* |
| *T. viride* | Chickpea seed treatment | Madhya Pradesh# |
| *T. harzianum* | Tomato seedling treatment | Meghalaya# |
| *T. viride* | Maize seed treatment | Meghalaya# |
| *T. viride* | Broccoli seed treatment | Sikkim# |
| *T. viride* | Nursery soil treatment | Sikkim# |
| *T. viride* | Tomato seed treatment | Sikkim# |
| *T. viride* | Capsicum seed treatment | Sikkim# |
| *T. viride* | Chilli seed treatment | Tamilnadu# |
| *T. harzianum* | Wheat seed treatment | Uttar Pradesh# |
| *Trichoderma* | Wheat seed treatment | Uttarakhand# |
| *Pseudomonas* | Wheat seed treatment | Uttarakhand# |
| *Pseudomonas + Trichoderma* | Pea seed treatment | Uttarakhand# |
| *Pseudomonas + Trichoderma* | Chickpea seed treatment | Uttarakhand# |
| *Trichoderma*<br>*Pseudomonas fluorescence* | Wheat seed treatment | Dharwad# |
| *Trichoderma* | Potato seed treatment | Dharwad# |

**National Centre of Organic Farming.*
*#Indian Institute of Farming Systems Research.*

**Table 2.**
*Seed treatment with promising* Trichoderma *spp. for organic agriculture.*

formulation could control root rot disease of citrus up to 80%, and under field conditions [42]. Some strains of *Pseudomonas* spp. and *Trichoderma* spp. found effective in controlling wilt of banana, caused by *F. oxysporum* f.sp. *cubense* (Foc) race 1 under field conditions [43]. Isolates of *Trichoderma* and *Aspergillus,* when applied in the field for the control of guava wilt disease caused by *F. oxysporum* f. sp. *psidii* and *F. solani*, could reduce disease incidence and promoted plant growth [38]. Antagonistic fungi *Aspergillus niger*, *T. harzianum,* and *Penicillium citrinum* were found effective for the management of the wilt disease of guava [44].

## 8. Management through foliar application of *Trichoderma* spp.

Foliar application of *Trichoderma* spp. is advisable for the management of airborne fungal and bacterial phytopathogens, such as species of *Alternaria*, *Curvularia*, *Fusarium*, *Colletotrichum*, *Pestalotiopsis*, *Pyricularia*, *Puccinia*, powdery and downy mildew pathogens of cereal, vegetable and other crops (**Table 3**). The potent candidates include *T. viride* and *T. harzianum*. Post-prune foliar application of *T. harzianum* and *T. viride* is a common practice in tea (*Camellia* sp) crop production in Darjeeling and the North East region of India [45, 46]. When a combination of *T. harzianum* and *Pseudomonas fluorescens* was sprayed before harvesting mango fruits, it had suppressed post-harvest fruit rot in dasheri mango [47]. Banana hands (cv. Grand Nain) when dipped in *T. asperellum* suspension and packed without ethylene absorbent

| Input | Crop disease | Adopted by |
|---|---|---|
| *Trichoderma* spp. | Anthracnose of chilli | Uttarakhand* |
| *Trichoderma* spp. | Rust, powdery mildew, and blight of pea | Uttarakhand# |
| *Trichoderma* spp. | Wilt and blight of chickpea | Uttarakhand# |
| *T. harzianum* | Black rust, brown rust, yellow rust, and leaf blight of wheat | Jharkhand# |
| *T. harzianum* or *T. viride* | Loose smut of wheat | Jharkhand# |

**National Centre of Organic Farming.*
*#Indian Institute of Farming Systems Research.*

**Table 3.**
*Foliar spray of* Trichoderma *spp. in organic agriculture.*

extended its shelf life up to 75 days at 13.5°C. There was no incidence of anthracnose and crown rot [41]. It was noted that *T. viride*, *T. harzianum*, and *T. asperellum* were potential antagonists for the management of *F. solani* and *Pestalotiopsis theae* causing dieback and grey leaf spot disease of tea [48, 49]. Foliar spray of *T. asperellum* and *T. atroviride* could manage the dieback disease of tea (*Camellia*sp) and enhance the vegetative growth in terms of more number of pluckable shoots [50].

## 9. Management through paste application of *Trichoderma* spp.

Application of *Trichoderma* paste (20% w/v) is generally done after severe pruning operations (rejuvenation prune and medium prune) in tea plantations to protect the plants from airborne pathogens. Such an application can be useful in horticultural crops in which pruning is done.

## 10. Important considerations for better performance of *Trichoderma* spp.

There are certain issues/concerns which should be taken into consideration for achieving desired benefits from the use of *Trichoderma* spp. Important concerns are—(1) reliable source of procurement, (2) assurance of fresh formulations, (3) avoidance of tank mixing with agrochemicals, (4) tank mixing with compatible bio formulations, (5) repeated application at the proper interval, (6) use of neat and clean separate sprayers for such formulations, (7) seeking technical advice from experts, (8) application during early morning and late evening hours, (9) procurement in well-ventilated store under lock and the key, and (10) quality assurance through certified agencies, such as IMO, Ecocert, Lecon, or any other approved agency.

## 11. Conclusion

*T. harzianum* and *T. viride* are efficient in controlling several plant diseases safely. When considering the interactions of *Trichoderma* fungi, it was found that these antagonistic fungi have an advantageous effect on plants. Stimulation of plant growth and yield takes place in this interaction and the advantageous effects are seen in the production of vitamins, the increased availability of biogenic elements (nitrogen, phosphorus), the mobilisation of nutrients from the soil and organic matter, and the enhanced intensity of mineral uptake and transport. Furthermore, *Trichoderma* fungi are capable of producing zeaxanthin and gibberellin, i.e., compounds accelerating seed germination. Managing plant diseases through these approaches could be safer for the agroecosystem, overall environment, soil health, and human health and would be the right step in sustaining crop production to meet the demand of the growing population.

## Acknowledgements

We, authors, are highly thankful to Dr. Azariah Babu, Deputy Director (Research), Tea Research Association, North Bengal Regional Research and Development Centre, Nagrakata, Jalpaiguri, West Bengal for critically going through the chapter.

## Conflict of interest

The authors declare no conflict of interest.

## Author details

Kishor Chand Kumhar[1*], Dalvinder Pal Singh[2] and Anil Kumar[3]

1 Chaudhary Charan Singh Haryana Agricultural University, Hisar, India

2 Forage Section, Department of Genetics and Plant Breeding, Hisar, India

3 Deendayal Upadhayay Center of Excellence for Organic Farming Chaudhary Charan Singh Haryana Agricultural University, Hisar, India

*Address all correspondence to: kishorkumarc786@hau.ac.in

## References


[1] Available from: https://www.fibl.org/fileadmin/documents/shop/1150-organic-world-2021.pdf [Accessed: February 6, 2022]

[2] Available from: https://apeda.gov.in/apedawebsite/organic/Organic_Products.htm [Accessed: February 6, 2022]

[3] Agrios GN. Plant Pathology. 5th ed. Gainesville, U.S.A.: Elsevier, Academic Press, University of Florida; 2005. p. 922

[4] Weindling R, Emerson OH. The isolation of toxic substance from the culture filterate of *Trichoderma*. Phytopathology. 1936;**26**:1068-1070

[5] Zaki O, Weekers F, Thonart P, Tesch E, Kuenemann P, Jacques P. Limiting factors of mycopesticide development. Biological Control. 2020;**144**:104220

[6] Bio-Pesticide Registrant | Directorate of Plant Protection, Quarantine & Storage | GOI (ppqs.gov.in) [Accessed: February 6, 2022]

[7] Harman GE. Myths and dogmas of biocontrol changes in perceptions derived on *Trichoderma harzianum* T-22. Plant Disease. 2000;**84**(4):377-393

[8] Howell CR. Mechanisms employed by *Trichoderma* species in the biological control of plant diseases; the history and evolution of current concepts. Plant Disease. 2003;**87**(1):4-10

[9] Benítez T, Rincón AM, Limón MC, Codon AC. Biocontrol mechanisms of *Trichoderma* strains. International Microbiology. 2004;**7**(4):249-260

[10] Smolińska U, Kowalska B, Oskiera M. The effectivity of strains in the protection of cucumber and lettuce against. Journal of Fruit and Ornamental Plant Research. 2007;**67**(1):81-93

[11] Kaczmarek J, Jędryczka M. Characterization of two coexisting pathogen populations of *Leptosphaeria* spp., the cause of stem canker of brassicas. Acta Agrobotanica. 2011;**64**(2):3-14

[12] Persoon CH. Disposita methodical fungorum. Romers Neues Mag Bot. 1794;**1**:81-128. [Accessed: February 6, 2022]

[13] Tulasne L, Tulasne R. De quelques Sphéries fungicoles, àpropos d'un mémoire de M. Antoine de Barysur les Nyctalis. Annales des Sciences Naturelles; Botanique. 1860;**13**:5-19

[14] Bisby GR. *Trichoderma viride* Pers. ex fries, and notes on Hypocrea. Transactions of the British Mycological Society. 1939;**23**(2):149-168

[15] Bissett J. A revision of the genus *Trichoderma*. II. Infrageneric classification. Canadian Journal of Botany. 1991;**69**(11):2357-2372

[16] Bissett J. A revision of the genus *Trichoderma*. III. Section Pachybasium. Canadian Journal of Botany. 1991; **69**(11):2373-2417

[17] Bissett J. A revision of the genus *Trichoderma*. IV. Additional notes on section Longibrachiatum. Canadian Journal of Botany. 1991;**69**(11):2418-2420

[18] Kullnig-Gradinger CM, Szakacs G, Kubicek CP. Phylogeny and evolution of the fungal genus *Trichoderma*: A multigene approach. Mycological Research. 2002;**106**(7):757-767

[19] Druzhinina IS, Kopchinsky AG, Kubicek CP. The first 100 *Trichoderma* species characterized by molecular data. Mycoscience. 2006;**47**(2):55-64

[20] Bissett J, Gams W, Jaklitsch W, Samuels GJ. Accepted *Trichoderma* names in the year 2015. IMA Fungus. 2015;**6**(2):263-295

[21] Herrera-Estrella A, Chet I. The biological control agent *Trichoderma* – From fundamentals to applications. In: Arora DK, Dekker M, editors. Fungal Biotechnology in Agricultural, Food and Environmental Applications. 21st ed. New York, USA: CRC Press; 2004. pp. 147-156

[22] Harman GE. Overview of mechanisms and uses of *Trichoderma* spp. Phytopathology. 2006;**96**(2):190-194

[23] Vinale F, Sivasithamparam K, Ghisalberti EL, Marra R, Woo SL, Lorito M. *Trichoderma* – Plant – Pathogen interactions. Soil Biology and Biochemistry. 2008;**40**(1):1-10

[24] Strakowska J, Błaszczyk L, Chełkowski J. The significance of cellulolytic enzymes produced by *Trichoderma* in opportunistic lifestyle of this fungus. Journal of Basic Microbiology. 2014;**54**(Suppl. 1):S2-S13. DOI: 10.1002/jobm.201300821

[25] Haran S, Schickler H, Oppenheim A, Chet I. Differential expression of *Trichoderma harzianum* chitinases during mycoparasitism. Phytopathology. 1996;**86**(9):980-985

[26] Jeleń H, Błaszczyk L, Chełkowski J, Rogowicz K, Strakowska J. Formation of 6-n-pentyl-2H-pyran-2-one (6-PAP) and other volatiles by different *Trichoderma* species. Mycological Progress. 2013;**13**(3):589-600. DOI: 10.1007/s11557-013-0942-2

[27] Lorito M, Hayes CK, Di Pietro A, Woo SL, Harman GE. Purification, characterization, and synergistic activity of a glucan 1,3-b-glucosidase and an N-acetyl-b-glucosaminidase from *Trichoderma harzianum*. Phytopathology. 1994;**84**(4):398-405

[28] Thrane C, Tronsmo A, Jensen DF. Endo-1,3-β-glucanase and cellulose from *Trichoderma harzianum*: Purification and partial characterization, induction of and biological activity against plant pathogenic *Pythium* spp. European Journal of Plant Pathology. 1997;**103**(4):331-344

[29] Viterbo A, Harel M, Chet I. Isolation of two aspartyl proteases from *Trichoderma asperellum* expressed during colonization of cucumber roots. FEMS Microbiology Letters. 2004;**238**(1):151-158

[30] Sharon E, Bar-Eyal M, Chet I, Herrera-Estrella A, Kleifeld O, Spiegel Y. Biological control of root-knot nematode *Meloidogyne javanica* by *Trichoderma harzianum*. Phytopathology. 2001;**91**(7):687-693

[31] Enari TM, Niku-Paavola ML. Enzymatic hydrolysis of cellulose: Is the current theory of the mechanisms of hydrolysis valid? Critical Reviews in Biotechnology. 1987;**5**(1):67-87

[32] Gams W, Bisset J. Morphology and identification of *Trichoderma*. In: Harman GE, Kubicek CP, editors. Trichoderma and Gliocladium. London, UK: Taylor & Francis; 1998. pp. 3-34

[33] Reino JL, Guerrero RF, Hernandez-Galan R, Collado IG. Secondary metabolites from species of the biocontrol agent *Trichoderma*. Phytochemistry Reviews. 2008;**7**(1):89-123

[34] Blaszczyk LMSKS, Siwulski M, Sobieralski K, Lisiecka J, Jedryczka M. *Trichoderma*

spp.–application and prospects for use in organic farming and industry. Journal of Plant Protection Research. 2014;**54**(4):309-317

[35] Corley DG, Miller-Wideman M, Durley RC. Isolation and structure of harzianum a: A new trichothecene from *Trichoderma harzianum*. Journal of Natural Products. 1994;**57**(3):422-425

[36] Monte E. Understanding *Trichoderma*: Between biotechnology and microbial ecology. International Microbiology. 2001;**4**(1):1-4

[37] Jeger MJ, Jeffries P, Elad Y, Xu XM. A generic theoretical model for biological control of foliar plant diseases. Journal of Theoretical Biology. 2009;**256**(2):201-214

[38] Gupta VK, Misra AK. Efficacy of bioagents against fusarium wilt of guava. Journal of Mycology and Plant Pathology. 2009;**39**(1):101

[39] Anonymous. DARE/ICAR Annual Report 2009-10. Crop Management. 2010. p. 48-63

[40] Anonymous. DARE/ICAR Annual Report 2013-14. Crop Management. 2014. pp. 56-67

[41] Available from: https://icar.org.in/files/DARE-ICARAnnualReport%20 2016-17English.pdf pp. 70. [Accessed: February 6, 2022]

[42] Available from: https://icar.org.in/files/DAREAnnual%20Report-2017-18_(English).pdf [Accessed: February 6, 2022]

[43] Bubici G, Kaushal M, Prigigallo MI, Gómez-Lama Cabanás C, Mercado-Blanco J. Biological control agents against fusarium wilt of banana. Frontiers in Microbiology. 2019;**10**:616

[44] Misra AK. Progressive steps in understanding and solving guava wilt-a national problem. Indian Phytopathology. 2017;**70**(1):1-11

[45] Anonymous. The Planters' Handbook. Digitised and Published by Tocklai Experimental Station. Jorhat – 785008, Assam: Tea Research Association; 2005. p. 131

[46] Barthakur BK, Dutta, P. Disease management in tea. In: Tea Field Management. A Compilation of Lecture Notes. B. K. Goswami. Jorhat, Assam, India: Tea Research Association, Tocklai Experimental Station; 2011. pp. 182-188

[47] Anonymous. DARE/ICAR Annual Report 2008-09. Crop Management. 2009. pp. 46-59

[48] Kumhar KC, Azariah B, Mitali B, Priyanka B, Tanima D. Biological and chemical control of *fusarium solani*, causing dieback disease of tea *Camellia sinensis* (L): An *in vitro* study. International Journal of Current Microbiology and Applied Sciences. 2015;**4**(8):955-963

[49] Kumhar KC, Babu A, Bordoloi M, Benarjee P, Rajbongshi H. Comparative bioefficacy of fungicides and *Trichoderma* spp. against *Pestalotiopsis theae*, causing grey blight in tea (*camellia* sp.): An *in vitro* study. International Journal of Current Research in Biosciences and Plant Biology. 2016;**3**(4):20-27

[50] Kumhar KC, Babu A, Arulmarianathan JP, Deka B, Bordoloi M, Rajbongshi H, et al. Role of beneficial fungi in managing diseases and insect pests of tea plantation. Egyptian Journal of Biological Pest Control. 2020;**30**(1):1-9

Chapter 5

# Using *Trichoderma* to Manage Sclerotia-Producing Phytopathogenic Fungi

*Jéssica Rembinski, Silvino I. Moreira, Jorge T. De Souza, Alan C.A. Souza, Adriano F. Dorigan, Eduardo Alves, Breno C.M. Juliatti and Fernando C. Julliati*

**Abstract**

Sclerotia are resistance structures that allow several soil-borne plant pathogens to survive for extended periods of time. The white mold disease, caused by *Sclerotinia sclerotiorum* and the stem rot in *Allium* spp., caused by *Stromatinia cepivora* are examples of destructive pathogens in which sclerotia are the central survival structure in their life cycle. In this chapter, we explore the information on the use of *Trichoderma* to manage sclerotia-producing pathogens in Brazil. There are 34 registered commercial products registered in Brazil, and most of them are recommended to manage sclerotia-producing fungi. The mechanisms of action of *Trichoderma* against these pathogens involve mainly mycoparasitism. The number of species employed as active ingredients of these commercial products is very limited, although many other species have shown a high potential against these pathogens. The white mold pathogen in soybean was taken as an example of field management, where the technical recommendations are detailed. This management involves other practices in addition to the application of *Trichoderma* in an integrated manner, and they are essential to manage this disease in the field in Brazil.

**Keywords:** biological control, white mold, white rot, resistance structure, plant disease management

## 1. Introduction

Sclerotia are survival melanized structures of different sizes and shapes depending on the fungal species and host plant. They remain viable in soil for long periods of time and are resistant to chemicals, adverse conditions, and biological degradation [1]. *Sclerotinia sclerotiorum*, *Rhizoctonia solani*, *Athelia rolfsii*, *Stromatinia cepivora*, and *Microphomina phaseolina* are among the most harmful sclerotia-producing phytopathogenic fungi in agriculture [2–6]. These pathogens are hard to manage and cause expanding losses in horticultural crops worldwide due to their survival capabilities associated with successive production of the same crop in the field and the lack of safe and efficient soil fumigation methods.

Biological control products formulated with *Trichoderma* have grown as the best method to manage sclerotia-producing pathogens, once chemicals in soil are often inefficient, too expensive, and too environmentally harmful. All *Trichoderma* species are able to utilize the cell content of other fungi as a source of nutrients [7–9]. However, certain species are more adapted to parasitize sclerotia of pathogenic fungi [2–6, 10–12]. These species are able to successfully invade, parasitize, and kill sclerotia in soil, meaning that the capacity to survive and compete in this environment is an absolute requirement. Although *Trichoderma* is one of the most commonly found genera in soil, not all species of the genus are well adapted to survive and thrive in soil [13]. The species adapted to soil need to be equipped with structures such as chlamydospores to ensure their long-term survival.

*Trichoderma* is a hyper diverse genus, with more than 350 species already described [14]. Only a small fraction of these species were developed into commercial products. This raises the questions which *Trichoderma* species have potential to be commercialized? Only the limited number of species that is well adapted to the soil environment and frequently are the active ingredients of formulations. Or is there any potential in the unexploited species?

In this chapter, we adopted a more practical view of the use of *Trichoderma* to control sclerotia-producing fungal plant pathogens. More emphasis is given on the mechanisms employed by *Trichoderma*, the species used in commercial formulations and the practical use of *Trichoderma* to control sclerotia producers in the field in Brazil, where products containing these fungi are used by farmers to manage white mold disease in soybean.

## 2. Plant pathogens capable of producing sclerotia

Sclerotia-producing fungi are very diverse, including saprotrophic, plant, animal and insect pathogens, mycoparasites, endophytes, insect symbionts, ecto- and ericoid-mycorrhizal fungi, and lichenicolous fungi [15]. However, they seem to be exclusively produced by fungi in the phyla Ascomycota and Basidiomycota that comprise the subkingdom Dikarya of the kingdom Fungi [15]. Sclerotia were documented in at least 85 fungal genera and 20 orders [15]. Outside of the kingdom fungi, sclerotia are relatively well studied in the slime molds or Myxomycetes [16], which belong to the kingdom Amoebozoa.

Sclerotia-producing fungi may be found in tropical and temperate regions, although sampling for these structures in natural environments is still scarce [15]. The structure and shape of fungal sclerotia are highly variable. While some are surrounded by much defined melanized ring encircling undifferentiated hyphae, others lack a distinct ring [17]. Some species, such as *Athelia rolfsii*, produce sclerotia with a round and smooth surface and dark color, whereas other fungi such as *Rhizoctonia solani* produce sclerotia with an irregular shape and lighter color. Some sclerotia are very small (sometimes smaller than 1 mm) such as the ones produced by *Macrophomina phaseolina*, *Stromatinia cepivora*, and *Verticillium dahliae*, whereas some fungal species, such as *Polyporus mylittae*, are able to produce giant sclerotia of up to 40 cm in diameter [18]. Fungi, such as *Claviceps purpurea*, *Sclerotinia sclerotiorum*, *Botrytis cinerea*, *Monilinia* spp., produce apothecia, which are sexual reproductive structures, directly on the sclerotia, whereas *Aspergillus flavus* produces cleistothecia inside the sclerotia [19].

Garlic and soybean are examples among the economically important crops where *Trichoderma* has been used to manage sclerotia-producing pathogens, at least under experimental conditions. Garlic cultivation in Brazil is done in the provinces of Goiás, Minas Gerais, Santa Catarina, and Rio Grande do Sul and it reached 118,000 tons in 2018 [20]. High losses have been induced by the white rot disease caused by the fungus *Stromatinia cepivora*, which is a pathogen specific to plants of the family Alliaceae [21]. *Stromatinia cepivora* belongs in the Ascomycota phylum, Helotiales order, and Sclerotiniaceae family [22]. The white rot caused in plants of the Alliaceae family presents well characterized symptoms, such as yellowing and death of infected leaves, owing to the root system damage caused by the pathogen. The aerial part of infected plants is easily detached from the soil. In garlic bulbs, the symptoms are soft rot of the tissues and white mycelial on the structural axis of infected plants and production of black microsclerotia on the bulbs [23]. The microsclerotia (ranging from 0.2 to 0.5 mm diameter) can remain dormant and viable in soil for more than 20 years in the absence of suitable host plants [24].

Soybean is an agricultural commodity traded not only in Brazil but is also exported [21]. In this crop, the white mold caused by *S. sclerotiorum* is considered the second most important disease after soybean rust [24, 25]. *Sclerotinia sclerotiorum*, which also belongs in the family Sclerotiniaceae, promotes the white mold or sclerotinia stem mold, leading to losses that can reach 70% in productivity. It occurs in approximately 30% of the Brazilian soybean producing area, which is currently more than 35 million ha [26]. The pathogen infects more than 600 plant species and is distributed worldwide. The sclerotia (from 0.5 to 10 mm diameter) are able to survive up to 12 years in soil (50, 51).

*Rhizoctonia solani* and *Athelia rolfsii* (syn. *Sclerotium rolfsii*) are two sclerotia-producing basidiomycetes from the families Ceratobasidiaceae and Atheliaceae, respectively, that cause rots in seeds, roots, stems, leaves, and fruits of a wide range of plant species (52, 53). These pathogens are commonly found in warm climates causing damping-offs, root rots, and wilts. The sclerotia are the only survival structures of these pathogens as they do not reproduce asexually and only rarely reproduce sexually.

These resistance structures are extremely difficult to destroy or inactivate by chemical methods, whereas physical methods such as solarization, inundation, and radiation are not feasible or too expensive to be adopted over large areas. In this context, biological control with antagonistic fungi of the genus *Trichoderma* is one of the best options available. The advantages of *Trichoderma* include its mycoparasitic capacity toward these structures, the excellent adaptation of certain species to the soil environment, the safety to humans and animals of the biocontrol species, and the relative ease of mass producing and formulating these agents.

## 3. Mechanisms employed by *Trichoderma* against sclerotia producers

The mechanisms of activity present in *Trichoderma* include mycoparasitism, antibiosis, competition, induction of resistance, and plant growth promotion [27–30]. Mycoparasitism is the main mechanism employed against sclerotia in soil [31], although the benefits observed in the field probably result from a combination of all mechanisms acting in concert. Mycoparasitism is defined as the ability of organisms to actively parasitize fungi and live at their expense [31].

Mycotrophy, a more inclusive term, may be defined as the ability of organisms to feed on either dead (passive mycotrophy, i.e., saprophytism) or on living fungi (active mycotrophy, i.e., mycoparasitism) [9]. This ability to feed on fungi, either dead or alive, was shown to be the ancestral form of nutrition in all *Trichoderma* species [13]. Although there are technical differences between mycotrophy and mycoparasitism, the latter term is the only one traditionally employed for *Trichoderma* in the literature, even in cases where there is no evidence that these fungi killed the host or prey.

Mycoparasitism by *Trichoderma* involves a sequence of events, including host localization, recognition, direct contact, coiling, formation of hook-shaped structures with appressorium function, penetration, folding, and development of parallel hyphae [29, 31–37]. It involves a combination of invasive hyphae with secondary metabolites and hydrolytic enzymes in most cases [38]. The wide range of *Trichoderma* secondary metabolites includes epipolythiodioxopiperazines (ETPs), peptaibols, pyrones, butenolides, pyridines, azaphilones, steroids, anthraquinones, lactones, trichothecenes, and harzianic acid [28]. These compounds can interfere with the metabolic activities of other microorganisms by inhibiting growth and sporulation, reducing spore germination, and weakening the sclerotia. Many *Trichoderma* species are strong producers of cellulases, chitinases and β-1,3-glucanases, proteases, and lipases, which act in concert with metabolites in the mycoparasitic activity of these fungi [31, 39–45].

Three distinct strategies of mycoparasitism were described for *Trichoderma*, which were supported by transcriptomic analyses [7]. These strategies are 1) passive or weak mycoparasitism, where species such as *T. reesei* have no capacity to stop the growth of fungi, but secrete cell wall degrading enzymes that slowly dissolve the mycelium of the host or prey; 2) strong mycoparasitism occurs in *T. atroviride* that actively and aggressively grows over the host and parasitizes it swiftly and produces proteases and glucanases; 3) mycoparasitism by lytic enzymes and toxic metabolites, which is observed in *T. virens*, that first produces metabolites such as gliovirin and gliotoxin that kill the host and later, the mycoparasite moves in and further produces lytic enzymes to digest the mycelium. Probably, strategies 2 and 3 will be more effective against sclerotia producers. There is solid evidence in the literature showing that *Trichoderma* behaves markedly differently in interactions with different fungal species or Oomycetes [9]. These differences are seen at the phenotypic level and at the gene expression level, where distinct mycoparasitism strategies are employed depending on the host or prey [9].

In this study, scanning electron microscopy observations of *in vitro* and in soil interactions between *Trichoderma* spp. and *S. sclereotiorum* (*Tr.* × *Ssc.*) and with *Stromatinia cepivora* (*Tr.* × *Sce.*) were done with scanning electron microscopy (SEM). Colonization of *S. sclerotiorum* sclerotia and *S. cepivora* microsclerotia by aerial mycelium of *Trichoderma* spp. was easily seen with the naked eye 7 days after the inoculation (**Figure 1A** and **I**). SEM analysis revealed the *Trichoderma* aerial mycelia-colonizing sclerotia of *Ssc.* and *Sce.* (**Figure 1B** and **J**). Conidia and conidiophores of *Trichoderma* spp. are produced on the surface of the sclerotia (**Figure 1C, D, M–O**). Cryo-fractures in both *Ssc.* sclerotia and *Sce.* microsclerotia evidenced the central medulla enclosed by the outer layer of rind cells free of *Trichoderma* colonization after 7 and 14 days after the incubation, respectively (**Figure 1B** and **K**). Aerial mycelia originated from the sclerotia were efficiently colonized by *Trichoderma* (**Figure 1E–H, M–P**). **Figure 1G** shows decaying hypha of *S. sclerotiorum*, indicating that cell wall-degrading enzymes acted on the pathogen.

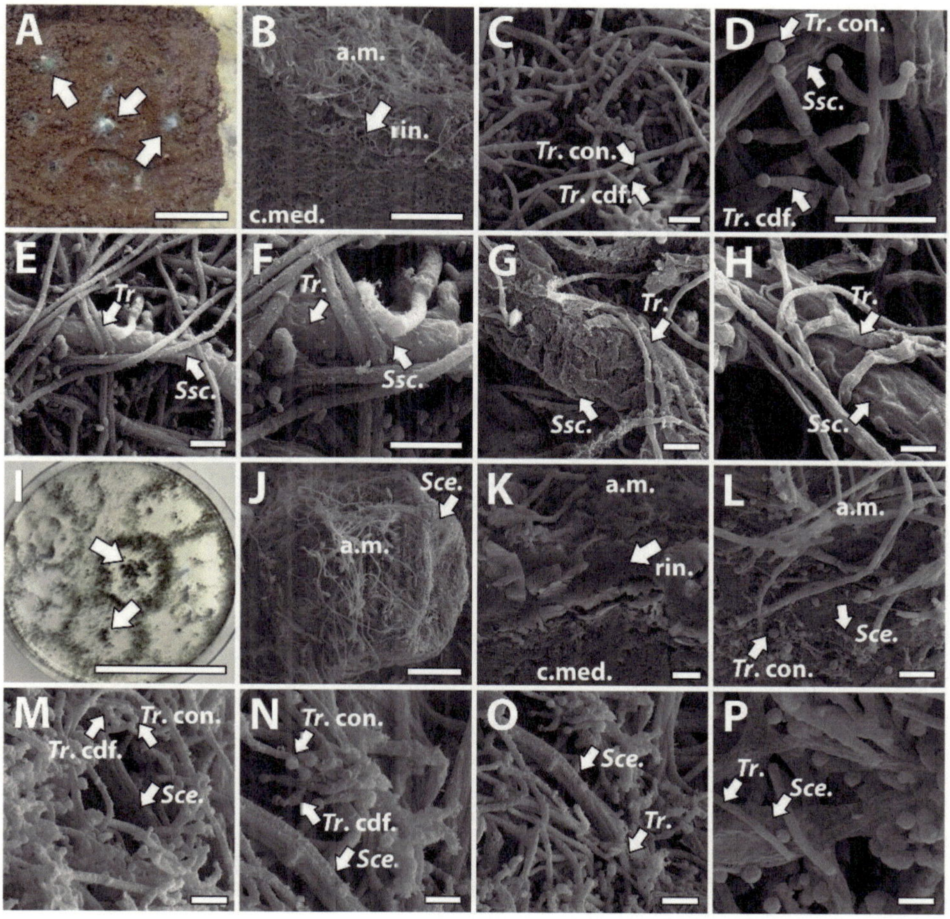


**Figure 1.**
*Scanning electron microscopy (SEM) observations of the interactions between* Trichoderma *and Sclerotinia sclereotiorum (Tr. × Ssc.) and with Stromatinia cepivora (Tr. × Sce.). A. Photography of Ssc. Sclerotia colonized by Tr. After seven days under 17°C above sterile soil.* B–H *and J–P: Scanning electron micrographs. B and K. Cryo-fractured Ssc. Sclerotia and Sce Microsclerotia evidencing the central medulla enclosed by the outer layer of rind cells, respectively. C–D. Trichoderma conidiophores producing conidia above Ssc. Sclerotia surface. E–H. Trichoderma-parasitizing Ssc. Hyphae. I. Photography of Sce. Microsclerotia colonized by Tr. After 14 days under 17°C above sterile soil. J–L. aerial mycelia of Sce. And Trichoderma above Sce. Microsclerotia. M–O. Trichoderma conidia and conidiophores above Sce. Microsclerotial surface. M–P.* Trichoderma *parasitizing Sce. Hyphae. The hyphae of Ssc. And Sce. were thicker than the hyphae of Trichoderma. a.m. = aerial mycelia; rin. = outer layer of rind cells in sclerotia; c.med. = central medulla enclosed by the rind cells in sclerotia; Tr.con. = Trichoderma conidia; Tr.cdf. = Trichoderma conidiophore; Tr. = Trichoderma; Ssc. = Sclerotinia sclerotiorum; Sce. = Stromatinia cepivora. Scale bars: A and I: 5 cm; B and J: 100 μm; C–H and K–P: 10 μm.*

## 4. Species of *Trichoderma* employed against sclerotia-producing fungi

In 2014, there were 177 *Trichoderma*-based fungicides commercially available in the world [46]. These products contained mainly *Trichoderma asperellum, T. hamatum, T. harzianum*, and *T. viride* as active ingredients and were recommended mainly for seed and soil treatments [46]. In Brazil, there are currently 34 formulated products with *Trichoderma* as active ingredients registered in the Ministry of Agriculture, Livestock and Food Supply (MAPA) (**Table 1**) [39, 47, 48]. These 34 products are based on four species: *T. harzianum*, *T. asperellum*, *T. koningiopsis*, and *T. stromaticum*.

| Commercial product | Active ingredient | Strain | Origin | Target pathogen |
|---|---|---|---|---|
| BF20.001 | *Trichoderma harzianum* + *Trichoderma asperellum* + *Bacillus amyloliquefaciens* | URM 8119 + URM 8120 + CCT 7901 | Ballagro Agro Tecnologia LTDA | *Colletotrichum lindemuthianum*, *Sclerotinia sclerotiorum*, *Rhizoctonia solani* |
| Bio-Hulk | *Trichoderma asperellum* | BV10 | Vittia Fertilizantes e Biológicos S.A. | *Rhizoctonia solani* |
| Pardella | *Trichoderma harzianum* + *Trichoderma asperellum* + *B. amyloliquefaciens* | URM 8119 + URM 8120 + CCT 7901 | Ballagro Agro Tecnologia LTDA | *Colletotrichum lindemuthianum*, *Sclerotinia sclerotiorum*, *Rhizoctonia solani* |
| Quality | *Trichoderma asperellum* | URM 5911 | Lallemand Soluções Agrobiológicas Ltda | *Fusarium solani f.sp. phaseoli*, *Sclerotinia sclerotiorum*, *Rhizoctonia solani* |
| Tanus | *Trichoderma harzianum* + *Trichoderma asperellum* + *B. amyloliquefaciens* | URM 8119 + URM 8120 + CCT 7901 | Biota Innovations Industria e Comercio de Bioprodutos Ltda | *Colletotrichum lindemuthianum*, *Sclerotinia sclerotiorum*, *Rhizoctonia solani* |
| Trichodermax EC | *Trichoderma asperellum* | T211 | Novozymes BioAg Produtos para Agricultura Ltda | *Fusarium solani f.sp. glycines*, *Sclerotinia sclerotiorum*, *Rhizoctonia solani* |
| Tricho-Guard | *Trichoderma asperellum* | BV10 | Vittia Fertilizantes e Biológicos S.A. | *Pratylenchus brachyurus*, *Rhizoctonia solani*, *Fusarium oxysporum*, *Sclerotinia sclerotiorum* |
| Tricho-Turbo | *Trichoderma asperellum* | BV10 | Vittia Fertilizantes e Biológicos S.A. | *Pratylenchus brachyurus*, *Rhizoctonia solani*, *Fusarium oxysporum*, *Sclerotinia sclerotiorum* |
| Tríppel | *Trichoderma asperellum* | CCT 2165 | Agrobiológica Sustentabilidade S.A. | *Rhizoctonia solani*, *Fusarium oxysporum f.sp. lycopersici* |
| Congregga | *Trichoderma asperellum* | CBMAI 1622 | Genica Inovação Biotecnologica S.A. | *Sclerotinia sclerotiorum* |
| Tricozak | *Trichoderma harzianum* + *Trichoderma asperellum* + *B. amyloliquefaciens* | URM 8119 + URM 8120 + CCT 7901 | Biota Innovations Industria e Comercio de Bioprodutos Ltda | *Colletotrichum lindemuthianum*, *Sclerotinia sclerotiorum*, *Rhizoctonia solani* |
| Organic WP | *Trichoderma asperellum* | URM 5911 | Lallemand Soluções Agrobiológicas Ltda | *Fusarium solani f.sp. phaseoli*, *Rhizoctonia solani* |
| Daytona | *Trichoderma harzianum* | ESALQ-1306 | Koppert do Brasil Holding Ltda. | *Fusarium solani f.sp. phaseoli*, *Rhizoctonia solani*, *Thielaviopsis paradoxa*, *Pratylenchus zeae*, *Sclerotinia sclerotiorum* |

| Commercial product | Active ingredient | Strain | Origin | Target pathogen |
|---|---|---|---|---|
| Ecotrich WP | *Trichoderma harzianum* | IBLF 006 | Ballagro Agro Tecnologia LTDA | *Sclerotinia sclerotiorum*, *Rhizoctonia solani*, *Macrophomina phaseolina*, *Sclerotium rolfsii* |
| Gaia Bio | *Trichoderma harzianum* | CCT 7589 | Simbiose Indústria e Comércio de Fertilizantes e Insumos Microbiológicos Ltda. | *Sclerotinia sclerotiorum*, *Rhizoctonia solani* |
| GreenControl | *Trichoderma harzianum* | SIMBI T5 (CCT 7589) | Simbiose Indústria e Comércio de Fertilizantes e Insumos Microbiológicos Ltda. | *Sclerotinia sclerotiorum*, *Fusarium solani f.sp. phaseoli* |
| Natucontrol | *Trichoderma harzianum* | BK-Th001 | Mezfer BR Soluções Agrícolas Ltda | *Sclerotinia sclerotiorum*, *Fusarium solani f.sp. phaseoli*, *Rhizoctonia solani* |
| Plant Protection | *Trichoderma harzianum* | CCT 7589 | Simbiose Indústria e Comércio de Fertilizantes e Insumos Microbiológicos Ltda. | *Sclerotinia sclerotiorum*, *Rhizoctonia solani* |
| Predatox | *Trichoderma harzianum* | IBLF 006 | Ballagro Agro Tecnologia LTDA | *Sclerotinia sclerotiorum*, *Rhizoctonia solani* |
| Shocker | *B. amyloliquefaciens* + *Trichoderma harzianum* + *B. amyloliquefaciens* | CPQBA 040-11DRM 01 + CPQBA 040-11DRM 09 + CPQBA 040-11DRM 04 | Agrivalle Brasil Industria e Comercio de Produtos Agricolas S.A. | *Sclerotinia sclerotiorum*, *Rhizoctonia solani* |
| Stimucontrol | *Trichoderma harzianum* | CCT 7589 | Simbiose Indústria e Comércio de Fertilizantes e Insumos Microbiológicos Ltda. | *Sclerotinia sclerotiorum*, *Rhizoctonia solani* |
| Stimucontrol Evolution | *Trichoderma harzianum* | CCT 7589 | Simbiose Indústria e Comércio de Fertilizantes e Insumos Microbiológicos Ltda. | *Sclerotinia sclerotiorum* |
| Torpeno | *B. amyloliquefaciens* + *B. amyloliquefaciens* + *Trichoderma harzianum* | CPQBA 040-11DRM 01 + CPQBA 040-11DRM 04 + CPQBA 040-11DRM 09 | Massen Produtos Biologicos S.A. | *Sclerotinia sclerotiorum*, *Rhizoctonia solani* |

| Commercial product | Active ingredient | Strain | Origin | Target pathogen |
|---|---|---|---|---|
| Trianum DS | *Trichoderma harzianum* | T-22 | Koppert do Brasil Holding Ltda. | *Fusarium oxysporum f.sp. lycopersici*, *Macrophomina phaseolina*, *Pratylenchus brachyurus*, *Sclerotinia sclerotiorum*, *Sclerotium rolfsii* |
| Trianum WG | *Trichoderma harzianum* | T-22 | Koppert do Brasil Holding Ltda. | *Macrophomina phaseolina*, *Pratylenchus brachyurus*, *Sclerotium rolfsii* |
| Trichodermaiz WP | *Trichoderma harzianum* | IB19/17 | Biocontrol Sistema de Controle Biológico Ltda | *Sclerotinia sclerotiorum* |
| Trichodermil SC 1306 | *Trichoderma harzianum* | ESALQ-1306 | Koppert do Brasil Holding Ltda. | *Fusarium solani f.sp. phaseoli*, *Rhizoctonia solani*, *Thielaviopsis paradoxa*, *Pratylenchus zeae*, *Sclerotinia sclerotiorum* |
| Trichodermil Super SC 1306 | *Trichoderma harzianum* | ESALQ-1306 | Koppert do Brasil Holding Ltda. | *Fusarium solani f.sp. phaseoli*, *Thielaviopsis paradoxa*, *Pratylenchus zeae*, *Sclerotinia sclerotiorum* |
| Trychonyd FR 25 | *Trichoderma harzianum* | CCT 6550 | TZ Biotec Ltda | *Sclerotinia sclerotiorum* |
| Walker | *Trichoderma harzianum* | T-22 | Koppert do Brasil Holding Ltda. | *Fusarium oxysporum f.sp. lycopersici*, *Sclerotinia sclerotiorum* |
| Rizoderma | *Trichoderma harzianum* | IBLF006 | Ballagro Agro Tecnologia LTDA | *Sclerotinia sclerotiorum*, *Rhizoctonia solani* |
| Tritter | *Trichoderma harzianum* | IBLF006 | Ballagro Agro Tecnologia LTDA | *Sclerotinia sclerotiorum*, *Rhizoctonia solani* |
| Lalnix Resist | *Trichoderma koningiopsis* | IBC 56/12 | Lallemand Soluções Agrobiológicas Ltda | *Meloidogyne incognita*, *Pratylenchus brachyurus*, *Heterodera glycines* |
| Tricovab | *Trichoderma stromaticum* | CEPLAC 3550 | CEPLAC | *Moniliophthora perniciosa* |

**Table 1.**
*Commercial products formulated with Trichoderma strains and registered in the Brazilian Ministry of Agriculture, Livestock and Food Supply [42].*

Most products are indicated for sclerotia-producing pathogens, for example, 26 products are recommended for *S. sclerotiorum*, 23 for *R. solani*, three for *Asclepias rolfsii*, and two for *Macrophomina phaseolina*. Most products are formulated with one strain and only six are combinations of strains. Some of the products are recommended to manage Oomycetes, nematodes, or for *Moniliophthora perniciosa*, the causative agent of cacao witches' broom disease (**Table 1**). No products are available for *S. cepivora*, even though potential *Trichoderma* strains are described in the literature [49].

Most commercial products based on *Trichoderma* are recommended for soil applications. Soil environments have few variations in temperature and humidity than the aerial parts of plants and these biocontrol agents show more potential in more stable niches. In 2020, there were more than 350 described species of *Trichoderma* in the world [14] and although only a limited number of species (approximately 30) appear to be well adapted to soil environments, the number of species used in commercial products is certainly under-represented. Additionally, it is possible that some of the species listed in **Table 1** are not identified correctly at the species level, as shown for members of the *T. harzianum* species complex [50].

Many species of *Trichoderma* other than the ones listed in **Table 1** were shown to have potential in the inactivation of sclerotia. *In vitro* assays performed by the first author of this chapter demonstrated the potential of different species of *Trichoderma* and eight undescribed species to colonize sclerotia of two pathogenic fungi (**Table 2**). Some of these strains were able to colonize up to 100% of the sclerotia of S. *cepivora* in soil (**Table 2**). The sclerotia of *S. sclerotiorum* appear to be more resistant to colonization than the ones produced by *S. cepivora*. Some of the novel strains were superior in comparison with a commercial product based on *T. harzianum*. These data underscore the potential of other than the species that are traditionally commercialized and novel *Trichoderma* species to be developed into commercial products to control sclerotia-producing plant pathogens. However, this potential has yet to be confirmed in field trials.

| Strain | Species* | Colonization (%) | |
|---|---|---|---|
| | | *S. cepivora* | *S. sclerotiorum* |
| Commercial strain | *T. harzianum* | 75 | 36 |
| TJ01 | *T. nordicum* | 100 | 100 |
| AAUT14 | *T. harzianum* | 69 | nd** |
| CX02TR07MTS | *T. harzianum* | 56 | nd |
| CX01TR06 | *T. breve* | 55 | nd |
| CX01TR02MTN | *T. breve* | 62 | nd |
| CX01TR04 | *T. breve* | 67 | nd |
| CX01TRCAM | *T. atroviride* | 100 | nd |
| AAUT7 | *T. orientale* | 71 | nd |
| AAUT11 | *T. orientale* | 89 | nd |
| CX01TR10 | *Tetramorium camerunense* | 80 | nd |
| CB1 | *Trichoderma* sp. 1 | 54 | nd |
| CB7 | *Trichoderma* sp. 1 | 38 | nd |
| CB8 | *Trichoderma* sp. 1 | 50 | nd |
| CX03TR10MTS | *Trichoderma* sp. 2 | 93 | nd |

| Strain | Species* | Colonization (%) | |
|---|---|---|---|
| | | *S. cepivora* | *S. sclerotiorum* |
| MON10A | *Trichoderma* sp. 3 | 100 | 20 |
| MON10B | *Trichoderma* sp. 3 | 100 | 38 |
| MON19 | *Trichoderma* sp. 3 | 100 | 25 |
| MON21 | *Trichoderma* sp. 3 | 100 | 40 |
| MTS13 | *Trichoderma* sp. 4 | 100 | 80 |
| CX01TR03 | *Trichoderma* sp. 5 | 82 | nd |
| CX01TR03MTN | *Trichoderma* sp. 5 | 87 | nd |
| CX01TR4.2MON | *Trichoderma* sp. 5 | 69 | nd |
| CX02TR18MTS | *Trichoderma* sp. 6 | 89 | nd |
| CX02TR05MTS | *Trichoderma* sp. 7 | 88 | nd |
| CX03TR04MTS | *Trichoderma* sp. 7 | 88 | nd |
| CX01TR4.1MON | *Trichoderma* sp. 8 | 90 | nd |
| CX01TR02MON | *Trichoderma* sp. 8 | 30 | nd |
| CX01TR1.2MON | *Trichoderma* sp. 8 | 73 | nd |

*Sclerotia of the pathogens were placed on the soil surface and sprayed with a suspension of* Trichoderma *conidia. The boxes containing the sclerotia were incubated at 17°C for* S. cepivora *and at 25°C for* S. sclerotiorum. *The evaluations were done at 7 and 14 days after the inoculations. The data are averages of two experiments with five replicates per experiment. * Species identification was done by sequencing the tef-1 fragment. * Not determined.*

**Table 2.**
*Colonization of sclerotia of* Stromatinia cepivora *and* Sclerotinia sclerotiorum *by strains of* Trichoderma.

## 5. Interaction with other beneficial microorganisms

The combination of more than one biocontrol agent is thought to be advantageous, but it depends on the individual strains compatibility [51]. Six out of the 34 registered products in Brazil are formulated with one or two *Trichoderma* and *Bacillus amyloliquefaciens* strains (**Table 1**). However, it is not known whether these microorganisms are compatible or not or if there is any synergism in their combination. Bacterial genera such as *Bacillus* and *Pseudomonas* are potential biocontrol agents of soil-borne pathogens due to the secretion of antibiotics and lytic enzymes in the rhizosphere of plants. Therefore, they are potential agents to be combined with *Trichoderma*, especially when they do not inhibit each other [51–55]. However, the compatibility of combinations needs to be evaluated with *in vitro* and *in planta* assays [56].

Interactions between *Trichoderma* and mycorrhyzae are sometimes antagonistic, such as with the ectomycorrhyzal basidiomycetous genus *Laccaria* spp., where there was clear inhibition of growth, colonization, and spore germination on both partners [57–59]. However, sometimes these interactions are synergistic, such as with *Glomus* spp. Although there was an increase in plant biomass in the interaction, microscopical observations clearly showed that *Trichoderma* was parasitizing this endomycorrhyzal fungus [60]. *Trichoderma* can parasitize the hyphae of the endomycorrhyzal fungus *Glomus irregulare* and gain entry into potato roots [61]. On the other hand, endomycorrhyzal species of *Rhizophaga* use *Trichoderma* to penetrate into the roots of non-host Brassicaceae, resulting in increased plant productivity [62].

The compatibility and synergism in interactions between *Trichoderma* and other beneficial microorganisms is so specific that they vary according to the strain of each partner and the host plant. Therefore, determining the outcome of these interactions is crucial for the successful field applications.

## 6. Field trials and uses in agriculture

Biological control of plant diseases is a reality in the agricultural world, since the abuse and inappropriate use of chemicals have led to major problems to the environment and human health. In view of sustainable agriculture, the use of chemical molecules is becoming unfeasible due to their high cost and toxicity. The use of products based on *Trichoderma* proved to be effective, especially against root pathogens able to produce resistance structures such as sclerotia [63–65]. The application of these microorganisms aiming to manage different plant diseases can be performed on seeds before planting, *via* foliar spraying, in the substrate, in the planting furrow or even in organic matter that will be incorporated before transplanting seedlings [66]. The form of application of products formulated with these microorganisms depends on the target to be controlled, the host crop, the environmental conditions, and the manufacturer's recommendations.

In Brazil, the control of white mold in soybeans is done by a combination of biological and chemical methods. This system will be used here as a case study to exemplify the application of *Trichoderma* in the field. White mold is the second most important soybean disease in Brazil and causes between 20 and 30% of losses on average, but under some conditions may reach 70–100% [26, 30, 67]. Approximately 10 million ha of soil is infested by this pathogen in Brazil out of total 35.9 million ha devoted to soybean cultivation in the country [67]. Approximately 5 million ha of soybean is currently treated with *Trichoderma*-based products in Brazil [67]. A common recommendation for white mold management is one spray application of *Trichoderma* at the vegetative stage V2 and another application at V4-V6 with concentrations varying from $10^9$ to $10^{11}$ CFU/ha, depending on the commercial product adopted. Spray application of *Trichoderma* should be done on overcast days with high soil humidity and mild temperatures. Since the levels of resistance in commercial cultivars are not satisfactory, the combination of other practices is desirable. No-tillage planting with mulch produced by *Urochloa ruziziensis* (Syn. *Brachiaria*) is highly recommended as it stimulates the sexual germination of sclerotia and at the same time functions a barrier for the spread of ascospores produced in apothecia [67, 68]. This mulch will also provide conducive conditions for the colonization of sclerotia and apothecia by *Trichoderma* [68, 69]. Another practice that must be adopted is the application of fungicides at the reproductive stage R1 and another application 15 days later. The most commonly used fungicides are fluazinam, thiophanate-methyl, procymidone, carbendazim, and trifloxystrobin [26]. Monitoring is essential for fine-tuning these recommendations to specific locations and environmental conditions. Some farmers may adopt the biological seed treatment with *Trichoderma* on top of the standard chemical seed treatment, as an additional measure to control damping-off, which is caused by many soil-borne pathogens, including *S. sclerotiorum*.

To verify the efficacy of the biological treatments, some of the manufactures of *Trichoderma* provide Petri plates with culture medium to farmers. Sclerotia should be collected from soil at the end of the cycle of soybean, but before harvest and plated to determine the percentage of sclerotia colonized by *Trichoderma*. When the

| Treatments | AUDPC | Incidence (%) | Sclerotia (g) | TGW (g) | Yield (kg/ha) |
|---|---|---|---|---|---|
| Control | 909.4 B | 25.5 B | 6.6 A | 127.6 B | 1942.5 C |
| *T. asperellum* SF04 (V4/V6) | 685.0 B | 9.2 A | 2.4A | 147.2 A | 2890.0 A |
| *T. harzianum* IBLF006 (V4/V6) | 766.9 B | 10.8 A | 5.8 A | 136.4 B | 2523.3 B |
| *T harzianum* ESALQ-1306. (V4/V6) | 651.2 B | 9.8 A | 2.2 A | 145.8 A | 2897.5 A |
| *Trichoderma* sp. Tricho 656 (V4/V6) | 656.9 B | 12.2 A | 4.6 A | 144.5 A | 2452.5 B |
| ESALQ-1306 (V4/V6) + Tioph. methyl (R1/R2) | 532.5 A | 10.5 A | 2.5 A | 146.1 A | 2765.0 A |
| Thiophanate methyl (R1/R2) | 658.1 B | 3.4 A | 3.4 A | 143.3 A | 2945.0 A |
| FLuazinam (R1/R2) | 440.6 A | 1.9 A | 1.9 A | 151.5 A | 3015.0 A |

*The experiment was done in the field with four replicates per treatment. Trichoderma was applied at a concentration of $10^9$ CFU/ha at the stages V4 and V6 and the fungicides at the stages R1 and R2. The area under the disease progress curve (AUDPC) was determined by integrating multiple measurements of disease severity; disease incidence was determined by measuring the percentage of infected plants per treatment; the number of sclerotia per ha was determined by separating them from seeds with sieves and weighing; TGW is the total grain weight and yield was determined by weighing. Means followed by different uppercase letters in the same columns are statistically different by the Scott-Knott test ($p \leq 0.05$).*

**Table 3.**
*Combination of Trichoderma and fungicides to control Sclerotinia stem rot in soybean.*

level of colonization of sclerotia in the plates is above 50%, the level of control is considered satisfactory. Although this level of control cannot be directly correlated with success, they serve to show that *Trichoderma* is present in the treated area at a relatively high rate.

Control of white mold is also dependent on the density of sclerotia in soil. Best results are obtained when the densities are 1–10 sclerotia/kg of soil [30]. Field experiments have shown that *Trichoderma* spp. as an exclusive control method is not sufficient to reduce the severity of white mold (**Table 3**). Fungicides normally have to be deployed to complement the activity of *Trichoderma*. However, in this experiment, where 1 year only was evaluated, two *Trichoderma* strains used alone were able to maintain the productivity at high levels even in the absence of fungicides. The use of *Trichoderma* in multiple years is expected to promote the build-up of inoculum in the soil and consequently decrease the levels of sclerotia to acceptable levels [70]. In the long run, the objectives are to maintain the inoculum in equilibrium and increase the plant growth and productivity.

## 7. Conclusion

In Brazil, 34 *Trichoderma*-based products are currently registered and most of them are recommended to control soil-borne pathogens that produce sclerotia as resistance structures. Relatively few species of the genus *Trichoderma* were developed into commercial formulations, despite the high number of publications that have shown the potential of many other species. *Trichoderma* spp. are widely used in Brazil to control white mold in soybean and its use is expected to increase in the near future as only 50% of the infected area is currently treated with these biocontrol agents. Besides providing partial control of white mold, these fungi can also increase plant growth and productivity coupled with a reduction in the use of chemical fungicides.

## Acknowledgements

JTS, EA, and FCJ acknowledge CNPq (National Council for Scientific and Technological Development) for his productivity scholarship.

## Conflict of interest

The authors declare no conflict of interest.

## Author details

Jéssica Rembinski[1], Silvino I. Moreira[1], Jorge T. De Souza[1*], Alan C.A. Souza[1], Adriano F. Dorigan[1], Eduardo Alves[1], Breno C.M. Juliatti[2] and Fernando C. Julliati[2*]

1 Plant Pathology Department, Federal University of Lavras, Lavras, MG, Brazil

2 Plant Protection Laboratory, Institute of Agricultural Sciences, Federal University of Uberlândia, Uberlândia, MG, Brazil

*Address all correspondence to: jorge.souza@ufla.br, juliatti@ufu.br and fernandocezar74@gmail.com

## References


[1] Smolinska U, Kowalska B. Biological control of the soil-borne fungal pathogen *Sclerotinia sclerotiorum*. Journal of Plant Pathology. 2018;**100**:1-12. DOI: 10.1007/s42161-018-0023-0

[2] Sá MNF, Lima JS, Jesus FN, Perez JO, Gava CAT. Effect of Bacillus sp. and Trichoderma sp. on mycelial growth of *Sclerotium rolfsii*. Acta Brasiliensis. 2019;**3**:79-81. DOI: 10.22571/2526-4338158

[3] Liang X, Rollins JA. Mechanisms of broad host range necrotrophic pathogenesis in *Sclerotinia sclerotiorum*. Phytopathology. 2018;**108**:1128-1140. DOI: 10.1094/PHYTO-06-018-0197-RVW

[4] McCaghey M, Willbur J, Smith DL, Kabbage M. The complexity of the *Sclerotinia sclerotiorum* pathosystem in soybean: Virulence factors, resistance mechanisms, and their exploitation to control Sclerotinia stem rot. Tropical Plant Pathology. 2019;**44**:12-22

[5] Bosamia TC, Dodia SM, Mishra GP, Ahmad S, Joshi B, Thirumalaisamy PP, et al. Unraveling the mechanisms of resistance to Sclerotium rolfsii in peanut (*Arachis hypogaea* L.) using comparative RNA-Seq analysis of resistant and susceptible genotypes. PLoS One. 2021;**15**(8):e0236823. DOI: 10.1371/journal. pone.0236823

[6] García VG, Onco MAP, Susan VR. Biology and systematics of the form genus Rhizoctonia. Spanish Journal of Agricultural Research. 2006;**4**(1):55-79

[7] Atanasova L, Le Crom S, Gruber S, Coulpier F, Seidl-Seiboth V, Kubicek CP, et al. Comparative transcriptomics reveals different strategies of *Trichoderma* mycoparasitism. BMC Genomics. 2013;**14**:121. Available from: http://www.biomedcentral.com/1471-2164/14/121

[8] Kubicek CP, Herrera-Estrella A, Seidl-Seiboth V. Comparative genome sequence analysis underscores mycoparasitism as the ancestral life style of *Trichoderma*. Genome Biology. 2021;**12**:1-15. DOI: 10.1186/gb-2011-12-4-r40

[9] Mukherjee PK, Mendoza-Mendoza A, Zeilinger S, Horwitz BA. Mycoparasitism as a mechanism of *Trichoderma*-mediated suppression of plant diseases. Fungal Biology Reviews. 2022;**39**:15e33. DOI: 10.1016/j.fbr.2021.11.004

[10] Abd-El-Moity TH. The use of *Trichodema* spp. to control soilborne plant pathogens in Egypt. In: Biological Control of Plant Diseases. Boston, MA: Springer; 1992. pp. 255-258

[11] Ararsa L, Thangavel S. Evaluation of arbuscular mycorrhizal fungi and Trichodema species for the control of onion white rot (*Sclerotium cepivorum* Berk). Journal of Plant Pathology & Microbiology. 2013;**4**:1000159. DOI: 10.4172/2157-7471.1000159

[12] Hussain WA, Elzaawely AA, Sheery NI, Ismail AA, El-Zahaby HM. Biological control of onion white rot disease caused by *Sclerotium cepivorum*. Environment, Biodiversity & Soil Security. 2017;**1**:101-107. DOI: 10.21608/jenvbs.2017.1547.1008

[13] Chaverri P, Samuels GJ. Evolution of habitat preference and nutrition mode in a cosmopolitan fungal genus with evidence of interkingdom host jumps and major shifts in ecology. Evolution. 2013;**67-10**:2823-2837. DOI: 10.1111/evo.12169

[14] Cai F, Druzhinina IS. In honor of John Bissett: Authoritative guidelines on molecular identification of *Trichoderma*. Fungal Diversity. 2021;**107**:1-69

[15] Smith ME, Henkel TW, Rollins JA. How many fungi make sclerotia? Fungal Ecology. 2014;**13**:211-220. DOI: 10.1016/j.funeco.2014.08.010

[16] Chet I, Henis Y. Sclerotial morphogenesis in fungi. Annual Review of Phytopathology. 1975;**13**:169-192

[17] Willetts HJ. Survival of fungal sclerotia under adverse environmental conditions. Biological Reviews of the Cambridge Philosophical Society. 1971;**46**:387-407

[18] Macfarlane TD, Kuo J, Hilton RN. Structure of the giant sclerotium of *Polyporus mylittae*. Transactions of the British Mycological Society. 1978;**71**:359-365

[19] Horn BW, Sorensen RB, Lamb MC, Sobolev VS, Olarte RA, Worthington CJ, et al. Sexual reproduction in *Aspergillus flavus* sclerotia naturally produced in corn. Phytopathology. 2014;**104**:75-85. DOI: 10.1094/PHYTO-05-13-0129-R

[20] IBGE. Instituto Brasileiro de Geografia de Estatística. Produção Agrícola Municipal–PAM [Internet]. 2020. Available from: https://www.ibge.gov.br/estatisticas/economicas/agricultura-e-pecuaria/9117-producao-agricola-municipal-culturastemporarias-e-permanentes.html?=&t=resultados. [Accessed: February 20, 2022]

[21] CONAB. Companhia Nacional de Abastecimento. Alho – Altos Preços, mas Clima Preocupa a Próxima Safra no Sul [Internet]. 2020. Available from: https://www.conab.gov.br/indicadores-da-agropecuaria/item/download/31689_569f5735ff790f58650ce6f2526323e4. [Accessed: February 20, 2022]

[22] Index Fungorum [Internet]. 2021. Available from: http://www.indexfungorum.org/names/names.asp. [Accessed: February 20, 2022]

[23] Domingos LB, Bontempo A, Lopes EA. Podridão-branca do alho e da cebola. In: Avanços Tecnológicos Aplicados à Pesquisa na Produção Vegetal. Universidade Federal de Viçosa [Internet]. 2015. Available from: https://www.researchgate.net/profile/Everaldo_Lopes/publication/283568389_Podridao-branca_do_alho_e_da_cebola/links/563f6ec008aec6f17ddb7ea5.pdf. [Accessed: February 20, 2022]

[24] Coley-Smith JR. White rot disease of *allium*: Problems of soil-borne diseases in microcosm. Plant Pathology. 1990;**39**:214-222

[25] Fuga CAG, Lopes EA, Vieira BS. Etiologia, epidemiologia e controle de doenças causadas por *Sclerotium rolfsii* e *S. cepivorum*. Revisão Anual de Patologia de Plantas. 2012;**20**:278-322

[26] Meyer MC, Campos HD, Godoy CV, Utiamada CM, Sato LN, Dias AR, et al. Eficiência de fungicidas para controle de mofo-branco (*Sclerotinia sclerotiorum*) em soja, na safra 2019/2020: resultados sumarizados dos experimentos cooperativos. Londrina: Embrapa; 2020. p. 10

[27] Adnan M, Islam W, Shabbir A, Khan KA, Ghramh HA, Huang Z, et al. Plant defense against fungal pathogens by antagonistic fungi with *Trichoderma* in focus. Microbial Pathogenesis. 2019;**129**:7-18. DOI: 10.1016/j.micpath.2019.01.042

[28] Khan RA, Najeeb S, Hussain S, Xie B, Li Y. Bioactive secondary metabolites from *Trichoderma* spp against phytopathogenic fungi. MDPI Microorganisms. 2020;**8**:817. DOI: 10.3390/microorganisms8060817

[29] Zhang F, Ge H, Zhang F, Guo N, Wang Y, Chen L, et al. Biocontrol potential of *Trichoderma harzianum* isolate T-aloe against *Sclerotinia sclerotiorum* in soybean. Plant Physiology and Biochemistry. 2016;**100**:64-74. DOI: 10.1016/j.plaphy.2015.12.017

[30] Juliatti FC, Rezende AA, Juliatti BC, Morais TP. *Trichoderma* as a biocontrol agent against *Sclerotinia* stem rot or white mold on soybeans in Brazil: Usage and technology. In: Shar MM, editor. *Trichoderma*: The Most Widely Used Fungicide. London, UK: IntechOpen; 2019. pp. 1-23. DOI: 10.5772/intechopen.84544

[31] Troian RF, Steindorff AS, Ramada MHS, Arruda W, Cirano JU. Mycoparasitism studies of *Trichoderma harzianum* against *Sclerotinia sclerotiorum*: Evaluation of antagonism and expression of cell wall-degrading enzymes genes. Biotechnology Letters. 2014;**36**:2095-2101. DOI: 10.1007/s10529-014-1583-5

[32] Abdullah MT, Ali NY, Suleman P. Biological control of *Sclerotinia sclerotiorum* (Lib.) de Bary with *Trichoderma harzianum* and *Bacillus amyloliquefaciens*. Crop Protection. 2008;**27**:1354-1359. DOI: 10.1016/j.cropro.2008.05.007

[33] Silva FF, Castro EM, Moreira SI, Ferreira TC, Lima AE. Alves. Emergência e análise ultraestrutural de plântulas de soja inoculadas com *Sclerotinia sclerotiorum* sob efeito da aplicação de *Trichoderma harzianum*. Summa Phytopathologica. 2017;**43**:41-45. DOI: 10.1590/0100-5405/2212

[34] Marcello CM. Avaliação da expressão e caracterização de uma exo-β-1,3-glucanase envolvida no mecanismo de micoparasitismo de *Trichoderma asperellum* [thesis]. Brasília: Universidade de Brasília; 2008

[35] Souza ACA, Sousa TP, Cortês MVB, Rodrigues FA, Silva GB, et al. Enzyme induced defense response in the suppression of rice leaf blast (*Magnaporthe oryzae*) by silicon fertilization and bioagents. International Journal of Research Studies in Bioscience. 2015;**3**:22-32

[36] Bozzola JJ, Russell LD. Electron Microscopy. Boston: Jones and Bartlett Publishers; 1999. p. 670

[37] Alves E, Lucas GC, Pozza EA, Carvalho Alves M. Scanning electron microscopy for fungal sample examination. In: Laboratory Protocols in Fungal Biology. New York: Springer; 2013. pp. 133-150

[38] Druzhinina IS, Seidl-Seiboth V, Herrera-Estrella A, Horwitz BA, Kenerley CM, Monte E, et al. *Trichoderma*: The genomics of opportunistic success. Nature Review Microbiology. 2011;**9**:749-759. DOI: 10.1038/nrmicro2637

[39] Bettiol W, Morandi MAB, Pinto ZV, Paula Junior TJ, Correa EB, Lucon CMM, et al, editors. Produtos Comerciais à Base de Agentes de Biocontrole de Doenças de Plantas. Jaguariúna: Embrapa Meio Ambiente; 2012. p. 155

[40] Dias PP. Controle biológico de fitopatógenos de solo por meio de isolados de fungos do gênero *Trichoderma* e sua contribuição no crescimento de plantas. 101 f. [thesis]. Seropédica: Universidade Federal Rural do Rio de Janeiro; 2011

[41] Amorim L, JAM R, Bergamin Filho A, editors. Manual de Fitopatologia vol. I–Princípios e Conceitos. 5th ed. Ouro Fino: Agronômica Ceres; 2018. p. 810

[42] Carvalho Filho MR. Relações filogenéticas, identificação e potencial de uso de isolados de *Trichoderma* no controle do mofo branco e como promotores de crescimento do feijoeiro [thesis]. Brasília: Universidade de Brasília; 2013

[43] Souza JR. Potencialidade de fungicida e agente biológico no controle de requeima do tomateiro [dissertation]. Vitória da Conquista: Universidade Estadual do Sudoeste da Bahia; 2013

[44] Melo IS. Potencialidades da utilização de *Trichoderma* spp. no controle biológico de doenças de plantas. In: Bettiol W, editor. Controle biológico de doenças de plantas. Jaguariúna: CNPDA/EMBRAPA; 1991. pp. 135-156

[45] Harman GE. The nature and application of biocontrol microbes II: *Trichoderma* spp. - Overview of mechanisms and uses of *Trichoderma* spp. Phytopathology. 2005;**96**:190-194. DOI: 10.1094/PHYTO-96-0190

[46] Woo SL, Ruocco M, Vinale F, Nigro M, Marra R, Lombardi N, et al. *Trichoderma*-based products and their widespread use in agriculture. The Open Mycology Journal. 2014;**8**:271-126

[47] MAPA. Ministério da Agricultura Pecuária e Abastecimento. Agrofit–Sistemas de Agrotóxicos Fitossanitários, Coordenação Geral de Agrotóxicos e Afins [Internet]. 2021. Available from: http://agrofit.agricultura.gov.br/agrofit_cons/principal_agrofit_cons. [Accessed: February 19, 2022]

[48] Lucon CMM, Chaves ALR, Bacilieri S, editors. *Trichoderma*: o que é, para que serve e como usar corretamente na lavoura. São Paulo: Instituto Biológico; 2014. p. 28

[49] Elshahawy IE, Saied N, Abd-El-Kareem F, Morsy A. Biocontrol of onion white rot by application of *Trichoderma* species formulated on wheat bran powder. Archives of Phytopathology and Plant Protection. 2017;**50**(3-4):150-166. DOI: 10.1080/03235408.2016.1276423

[50] Chaverri P, Branco-Rocha F, Jaklitsch W, Gazis R, Degenkolb T, Samuels GJ. Systematics of the *Trichoderma harzianum* species complex and the re-identification of commercial biocontrol strains. Mycologia. 2015;**107**(3):558-590. DOI: 10.3852/14-147

[51] Barbosa LO, Lima JS, Magalhaes VC, Gava CAT, Soares ACF, Marbach PAS, et al. Compatibility and combination of selected bacterial antagonists in the biocontrol of sisal bole rot disease. BioControl. 2018;**63**(4):595-605. DOI: 10.1007/s10526-018-9872-x

[52] Mishra J, Arora NK. Secondary metabolites of fluorescent pseudomonads in biocontrol of phytopathogens for sustainable agriculture. Applied Soil Ecology. 2017;**125**:35-45. DOI: 10.1016/j.apsoil.2017.12.004

[53] Pawana G, Badami K, Djunaedy A. Growth and reproduction of *Trichoderma* sp. in with presence *Bacillus* sp. or fluorescent pseudomonad. Journal of Physics: Conference Series. 2021;**1899**:1-7. DOI: 10.1088/1742-6596/1899/1/012021

[54] Pan S, Jash S. Variability in biocontrol potential and microbial interaction of *Trichoderma* spp. with soil inhabiting antagonistic bacteria *Pseudomonas fluorescens*. Indian Phytopathology. 2010;**63**:158-164

[55] Braga AF. Interaction of *Trichoderma Asperellum* and Biological Control of *Bacillus* spp. Used in Soybean Diseases [Dissertation]. Instituto Federal Goiano: Rio Verde; 2021

[56] Cruz-Magalhães V, Guimarães RA, Silva JCP, Faria AF, Pedroso MP, Campos VP, et al. The combination of two *Bacillus* strains suppresses *Meloidogyne incognita* and fungal pathogens, but does not enhance plant growth. Pest Managment Science. 2021;**78**(2):722-732. DOI: 10.1002/ps.6685

[57] Summerbell RC. The inhibitory effect of *Trichoderma* species and other soil microfungi on formation of mycorrhiza by *Laccaria bicolor in vitro*. New Phytology. 1987;**105**:437e448. DOI: 10.1111/j.1469-8137.1987.tb00881.x

[58] Werner A, Zadworny M, Idzikowska K. Interaction between *Laccaria laccata* and *Trichoderma virens* in co-culture and in the rhizosphere of *Pinus sylvestris* grown *in vitro*. Mycorrhiza. 2002;**12**:139e145. DOI: 10.1007/s00572-002-0159-8

[59] Guo Y, Ghirardo A, Weber B. *Trichoderma* species differ in their volatile profiles and in antagonism toward ectomycorrhiza *Laccaria bicolor*. Frontiers in Microbiology. 2019;**10**:891. DOI: 10.3389/fmicb.2019.00891

[60] Rousseau A, Benhamou N, Chet I, Piche Y. Mycoparasitism of the extramatrical phase of *Glomus intraradices* by *Trichoderma harzianum*. Phytopathology. 1996;**86**:434e443. DOI: 10.1094/Phyto-86-434

[61] De Jaeger N, Declerck S, De La Providencia IE. Mycoparasitism of arbuscular mycorrhizal fungi: A pathway for the entry of saprotrophic fungi into roots. FEMS Microbiology Ecology. 2010;**73**:312e322. DOI: 10.1111/j.1574-6941.2010.00903.x

[62] Poveda J, Hermosa R, Monte E, Nicolas C. *Trichoderma harzianum* favors the access of arbuscular mycorrhizal fungi to non-host Brassicaceae roots and increases plant productivity. Scientific Reports. 2019;**9**:1-11. DOI: 10.1038/s41598-019-48269-z

[63] Souza ACA, Filippi MCC, Nascente AS, Prabhu AS, Alves E. Silicon rates and beneficial microorganism on blast suppression and productivity of upland rice. Journal of Plant Science and Phytopathology. 2021;**5**:20-27. DOI: 10.29328/journal.jpsp.1001057

[64] Caixeta GJ, Nascimento LM, Filippi MCC, Gonçalves FJ, Souza ACA. Use of bioagents in the suppression of *Macrophomina phaseolina* in the common bean culture. Ipê Agronomic Journal. 2021;**5**:1-11

[65] Silva NP, Filippi MCC, Gonçalves FJ, Souza ACA. Biocontrol of *Macrophomina phaseolina* and promoting growth in sunflower culture with the use of *Trichoderma* sp. and Rhyzobacteria. Ipê Agronomic Journal. 2021;**1**:1-14

[66] Moscoso MAP, Ramírez IJN, Angarita CC, Rojas ET. In vitro biocontrol activity of macrohongs against different phytopathogenic fungi. Acta Biológica Colombiana. 2020, 2020;**25**:265-279. DOI: 10.15446/abc.v25n2.75303

[67] Meyer MC, Campos HD, Godoy CV, Utiamada CM. Ensaios cooperativos de controle biológico de mofo branco na cultura da soja—safras 2012 a 2015. Londrina: Embrapa Soja; 2016. p. 49

[68] Görgen CA, Civardi E, Perreto E, Carneiro LC, Silveira Neto AN, Ragagnin V, et al. Controle de *Sclerotinia sclerotiorum* com o manejo de *Brachiaria ruziziensis* e aplicação de *Trichoderma harzianum*. Circular Técnica 81: Embrapa; 2008. p. 4

[69] Silva JC, Meyer MC, Mazaro SM. *Trichoderma*: uso na agricultura. Brasília, DF: Embrapa; 2019. p. 536

[70] Jaccoud Filho DS, Henneberg L, Grabicoski EMG, editors. Mofo branco. Ponta Grossa: Todapalavra; 2017. p. 520